International Conference on Plant Synthetic Biology and Bioengineering 2016

Miami Beach, Florida, USA
16 - 18 December 2016

ISBN: 978-1-5108-4166-6

Printed from e-media with permission by:

Curran Associates, Inc.
57 Morehouse Lane
Red Hook, NY 12571

Some format issues inherent in the e-media version may also appear in this print version.

Printed by Curran Associates, Inc. (2017)

For permission requests, please contact AIChE
at the address below.

AIChE
120 Wall Street, FL 23
New York, NY 10005-4020

Phone: (800) 242-4363
Fax: (203) 775-5177

www.aiche.org

Additional copies of this publication are available from:

Curran Associates, Inc.
57 Morehouse Lane
Red Hook, NY 12571 USA
Phone: 845-758-0400
Fax: 845-758-2633
Email: curran@proceedings.com
Web: www.proceedings.com

GENERAL SUBMISSIONS

12:00 AM
Mining Bacterial Genomes for a 'Hack-Proof' Thiamin Biosynthesis Enzyme 1

- Andrew D. Hanson
- Antje M. Thamm
- Svetlana Y. Gerdes
- Valerie de Crecy-Lagard
- Steven D. Bruner
- Gengnan Li

12:25 AM
A Robust Gene Stacking Method Utilizing Yeast Assembly for Plant Synthetic Biology 3

- Patrick Shih

12:50 AM
Design and Implementation of Genetic Circuits for Complex Biological Functions in Plants 4

- Mauricio Antunes
- June I. Medford
- Kevin Morey
- Katherine Schaumberg

1:15 AM
Development of a Two-Component System for Signal Propagation in Plants 6

- Tom Schreiber
- Alain Tissier

1:40 AM
Determining the Principles behind Tissue-Specific Gene Expression 8

- Molly Megraw

12:00 AM
Rational Design and Biological Complexity: Quantitative Measurement and Prediction of Synthetic Circuits Using Plant Protoplasts 9

- June I. Medford
- Katherine Schaumberg
- Mauricio Antunes

- Tessema Kassaw
- Christopher Zalewski

12:25 AM
Computational Modeling of Cytoskeletal and Cell Wall Systems during Cell Morphogenesis 11

- Dan Szymanski

12:50 AM
Development and Application of Synthetic Biology Tools to Modulate Decarboxylation in Arabidopsis 13

- Henrique C. De Paoli
- Xiaohan Yang
- Gerald Tuskan
- Timothy J. Tschaplinski
- Robert Hettich
- Nancy L. Engle
- Anne Borland
- Paul E. Abraham
- Yaojin Sun

1:15 AM
Reprogramming the plant form: Using synthetic signalling to study and re-engineer development 15

- Arjun Khakhar
- Andrew Lemmex
- Eric Klavins
- Jennifer Nemhauser

1:40 AM
Synthetic Programming of Combinatorial Gene Regulation in Plant Development 17

- Edith Pierre-Jerome
- Philip Benfey

2:05 AM
Evolvability and Functional Divergence in the Auxin-Signaling F-Box Family 18

- R. Clay Wright
- Mollye L. Zahler
- Stacey R. Gerben
- Jennifer Nemhauser

12:00 AM
Production of Halogenated Indigo Precursors in Metabolically Engineered Tobacco Plants 20

- Heribert Warzecha
- Sabine Fräbel

12:25 AM
Red Biotechnology Potential of Taxol-Producing Endophytic Fungus Lasodiplodia Theobromea Isolated from a Non-Taxol-Producing Plant 22

- Balendra Sah
- Kamalraj S.
- Jayabaskaran C.

12:50 AM
Development of a High-Throughput Physcomitrella patens Transformation System for Biosynthetic Applications 24

- Antony Evans
- Jihyun Moon
- Jamey Kain
- James Anderson-Furgeson

1:15 AM
A Synthetic Biology Yeast Platform for the Biosynthesis of Plant Phenolic Diterpenes 26

- Ulschan Scheler
- Alain Tissier

12:00 AM
Cutting-Edge Methods and Tools for Plant Genome Editing 28

- Tomas Cermak
- Shaun Curtin
- Javier Gil-Humanes
- Radim Cegan
- Colby Starker
- Thomas Kono
- Eva Konecna
- Jade Mathre
- Becca Greenstein
- Robert Stupar
- Daniel Voytas

12:25 AM

Crispr-Cas9 or Talen Mediated Genome Editing in Sugarcane 30

- Fredy Altpeter
- Tufan Oz
- Baskaran Kannan
- Ratna Karan
- Je Hyeong Jung
- Aldo Merotto

12:50 AM
Nanoparticles As Biomolecular Cargo Transporters in Plants and Plastids 32

- Gozde Sultan Demirer
- Markita Landry

12:25 AM
Functional Analyses of Cotton (Gossypium hirsutum L.) Immature Fiber (im) Mutant Whose Phenotype Grown Under Normal Conditions Is Similar to Wild Type Cotton Grown Under Severe Stress 34

- Hee-Jin Kim
- Christopher Delhom
- Yongliang Liu
- James Rodgers
- David Fang
- Seong H. Kim

12:50 AM
Insight into the Plant Stress Adaptation Strategies during Tripartite Interaction of Plant-Pathogen and Nanoparticles Using Comparative Proteomics Approach 36

- Madhuree Kumari
- Lalit Agarwal
- Shipra Pandey
- Aradhana Mishra
- Chandra Nautiyal

1:15 AM
Fruit Crop Engineering with Igt Family Genes 38

- Jessica M. Guseman
- Kevin Webb
- Courtney A Hollender
- Doug Raines
- Chinnathambi Srinivasan
- Chris Dardick

12:00 AM
Engineering for Robust Photorespiratory Bypass to Improve Photosynthesis and Crop Production Under Field Conditions 40

- Paul F South
- Amanda P Cavanagh
- Donald Ort

12:25 AM
Redesigning Photosynthesis to Sustainably Meet Global Food and Demand of Bioenergy 42

- Bilal Ali Khan

12:50 AM
Photosynthetic Antenna Engineering to Improve Crop Yields 44

- Anastasios Melis

1:15 AM
Agave and Kalanchoe Omics for Synthetic Biology to Enhance Water Use Efficiency and Drought Tolerance in Plants 46

- Xiaohan Yang
- Rongbin Hu
- Hengfu Yin
- Degao Liu
- Jeremy Schmutz
- Jerry Jenkins
- Shengqiang Shu
- David Goodstein
- Paul E. Abraham
- Robert Hettich
- John C. Cushman
- Anne Borland
- Timothy J. Tschaplinski
- Gerald Tuskan

1:40 AM
Intracellular Spectral Recompositioning of Light Increases Photosynthetic Efficiency and Reduces Light Associated Stress in a Versatile Model Diatom 48

- Weiqi Fu
- Amphun Chaiboonchoe
- Basel Khraiwesh
- Mehar Sultana
- Ashish Jaiswal

- Kenan Jijakli
- David Nelson
- Kourosh Salehi-Ashtiani

12:00 AM
Fusion of Ferredoxin to Cytochrome P450 Enables Light-Driven Biosynthesis 50

- Poul Erik Jensen

12:00 AM
Genes Stacking of Multiple Traits to Improve Biomass for Biofuels 52

- Aymerick Eudes
- Khanh Vuu
- Patrick Shih
- Dominique Loqué
- Aude Aznar
- Camille Chalvin
- Mi Yeon Lee
- Vibe M. Gondolf
- Berit Ebert
- Devon Birsdeye
- Tomas Laursen
- Henrik V. Scheller

12:25 AM
Engineering Functionally Specialized Monolignol 4-O-Methyltransferases to Alter Lignin Structure in Energy Crop 54

- Chang-Jun Liu

12:50 AM
Metabolic Engineering Converts Sugarcane into an Oil Producing Crop 56

- Fredy Altpeter
- Saroj Parajuli
- Ratna Karan
- Baskaran Kannan
- Georgina Sanahuja
- Hui Liu
- John Shanklin

1:15 AM
Sugar Transport By Sweets: From Structures to Biosensors 58

- Lily Cheung

- Wolf Frommer
- Taylor Chavez

1:40 AM
Increasing the C6/C5 Sugar Ratio in Bioenergy Crops By Modulating Nucleotide Sugar Transporters and Glycosyltransferases 59

- Henrik V. Scheller
- Aude Aznar
- Berit Ebert
- Devon Birdseye
- Dominique Loqué
- Camille Chalvin
- Tomas Laursen
- Andrew G. Brandon
- Carsten Rautengarten
- Joshua Heazlewood
- Jingwei Yan

POSTER SUBMISISSIONS

5:20 PM
A Toolkit of Programmable Salicylic Acid Regulated Transcription Factors to Engineer Plant Immune Responses 61

- Orlando de Lange
- Arjun Khakhar
- Jennifer Nemhauser
- Eric Klavins

5:20 PM
Bicistronic Expression of Gibberellin 20-Oxidase and Secondary Wall Biosynthetic MYB Transcription Factor Improves Both Quality and Quantity of Woody Biomass Production in Poplar 63

- Jae-Heung Ko
- Jin-Seong Cho
- Eung-Jun Park
- Young-Im Choi

5:20 PM
Bioproduction of Muconic Acid in Plants 65

- Khanh Vuu
- Patrick Shih
- Aymerick Eudes
- Dominique Loque

5:20 PM
Bistable Genetic Switch Design in Plant System 66

- Tessema Kassaw
- Christopher Zalewski
- Katherine Schaumberg
- Mauricio Antunes
- Ashok Prasad
- June I. Medford

5:20 PM
Characterization of Arabidopsis Mutants with Minimal Level of Centromeric Histone: Impact on Chromosome Segregation and Induction of Haploid Plants 68

- Claudio Capitao
- Svetlana Akimcheva
- Sorin Tanasa
- Petra Bulankova
- Vivek Raxwal
- Ortrun Mittelsten Scheid
- Karel Riha

5:20 PM
Comparative Transcriptome Analysis of Qualea Grandiflora mart. Plants to Investigate the Role of Aluminum in This Cerrado Species Metabolism 70

- Renata Cristina Costa e Silva

5:20 PM
Comprehensive Analysis of Heavy Metal-Associated Domain Containing Genes Using Nitric Oxide-Mediated Transcriptomic Analysis in Arabidopsis 72

- Qari Muhammad Imran
- Noreen Falak
- Adil Hussain
- Bong-Gyu Mun
- Byung-Wook Yun

5:20 PM
Crop-Biorefinery System for Low-Cost Advanced Renewable Fuel and Chemicals 74

- Ai Oikawa

5:20 PM
Design, Synthesis, and Testing Toward a 57-Codon Genome 76

- Nili Ostrov

5:20 PM
Dominant Suppression of Xylan Biosynthesis Using IRX10 78

- Andrew G. Brandon

5:20 PM
Engineering of Crassulacean Acid Metabolism (CAM) Modules into C3 Species Via Synthetic Biology Approach 80

- Rongbin Hu
- Degao Liu
- Anne Borland
- Gerald Tuskan
- Xiaohan Yang

5:20 PM
Genome-Wide Study on Cultivar-Specific and Stripe Rust Responsive Mirnas in Triticum aestivum 82

- Sowmya Ramachandran

5:20 PM
In Situ engineering of Bacterial Communities 84

- Jennifer A.N. Brophy
- Alan D. Grossman

5:20 PM
Inter-Specific Differences in Growth, Leaf Phenology in Seedlings of Cork Oak (Quercus suber L.), Zeen Oak (Quercus canariensis Willd.) and Their Hybrid Afares Oak (Quercus afares Pomel) Grown in the Nursery 85

- Sameh Mhamdi

5:20 PM

Lectin Receptor Protein Kinase Is Induced in Plant Roots in Response to the Endophytic Fungus, Piriformospora indica 87

- Nivedita Lal
- M Z Abdin
- Praveen Verma
- K C Upadhyaya

5:20 PM
Lipid Droplets As a New Platform for Protein Expression in Chlamydomonas Reinhardtii 89

- Poul Erik Jensen
- Agnieszka Zygadlo Nielsen
- Kamil Bakowski

5:20 PM
Metabolic Profile of Qualea Grandiflora mart. (Vochysiaceae) Reveals the Importance of Aluminium in This Plant Metabolism 91

- Jessica R. Melo

5:20 PM
Phosphatase from the Haloacid Dehalogenase (HAD) Superfamily Catalyze the Elusive Dephosphorylation Step of Riboflavin Biosynthesis in Arabidopsis thaliana Chloroplasts 94

- Na Sa

5:20 PM
Plant Foundry: Biopolymers, Biopharmaceuticals and Medically Effective Secondary Metabolites (Vaccines, antibodies, and metabolites) 96

- Ethenia Scott

5:20 PM
Plant Pathway Discovery and Bioengineering of the Most Popular Flavor, Vanillin 98

- Nethaji Gallage

5:20 PM
Precision Genome Editing in Sugarcane 101

- Tufan Oz
- Ratna Karan
- Fredy Altpeter

5:20 PM

Regulation of Stomatal Movement Using Synthetic Biology Approach 102

- Degao Liu
- Rongbin Hu
- Anne Borland
- Xiaohan Yang

5:20 PM
Stable Expression of the Fungal Effector BEC1019 in Transgenic Barley to Evaluate Host-Pathogen Interactions 104

- Tufan Oz
- Weihui Xu
- Roger Wise
- Fredy Altpeter

5:20 PM
SynCellBio: An Inquiry-Based Research Course Exploring Synthetic Biology 106

- Britney Moss

5:20 PM
Synthetic Biology of N2-Fixing Cyanobacteria for Photosynthetic Production of Perfumed Linalool from Air and Water 108

- Liping Gu
- Charles Halfmann
- Trevor VanDenTop
- Nathanael Braselton
- Ruanbao Zhou
- William Gibbons

5:20 PM
Talen Mediated Multi-Allelic Mutagenesis of COMT Improves Saccharification Efficiency from Lignocellulosic Biomass of Field Grown Sugarcane 110

- Baskaran Kannan
- Je Hyeong Jung
- Fredy Altpeter

5:20 PM
Towards Design of Endophytic Bacteria That Can Deliver Fixed Nitrogen to Cereal Crops 112

- Min-Hyung Ryu
- Miryoung Song

- Kevin Garcia
- Shanmugam Rajasekar
- Jean-Michel Ané

Mining Bacterial Genomes for a 'Hack-Proof' Thiamin Biosynthesis Enzyme

- Conference: International Conference on Plant Synthetic Biology and Bioengineering
- Year: 2016
- Proceeding: International Conference on Plant Synthetic Biology and Bioengineering
- Group: General Submissions
- Session:

  Rational Design and mining of natural systems for plant synthetic biology components

- Time:

  Friday, December 16, 2016 - 12:00am-12:25am

Authors:
Hanson, A. D., University of Florida
Thamm, A. M., University of Florida
Gerdes, S. Y., Fellowship for Interpretation of Genomes
de Crecy-Lagard, V.
Bruner, S. D.
Li, G.
Vitamins and their precursors are key currencies exchanged between within and between plants and their microbiomes. These vitamin transactions can be manipulated to favor or disfavor particular organisms. An example is the natural antibiotic bacimethrin, an analog of the thiamin precursor hydroxymethylpyrimidine (HMP) whose toxicity depends on replacing HMP in thiamin synthesis, i.e. on hacking into the thiamin pathway. Bacimethrin biosynthesis could in principle be installed in plants â€“ but the plant platform would need first to be made bacimethrin-resistant (hack-proof). Bacimethrin enters the thiamin pathway via ThiD, a kinase present in plants and most bacteria that

converts the biosynthetic intermediate HMP monophosphate to the diphosphate but also has a salvage activity that converts HMP (or bacimethrin) to the monophosphate. We therefore used comparative genomics to prospect for enzymes lacking the salvage activity. We uncovered a small protein of unknown function, often fused to the thiamin synthesis enzyme ThiE, that we predicted could replace the biosynthetic activity of ThiD but not its salvage activity. Genetic and biochemical analyses confirmed that this protein, named ThiD2, catalyzes the phosphorylation of HMP monophosphate, but not of HMP, bacimethrin, or other toxic HMP analogs. As a strictly monofunctional HMP monophosphate kinase, ThiD2 eliminates an inbuilt vulnerability of canonical ThiD.

A Robust Gene Stacking Method Utilizing Yeast Assembly for Plant Synthetic Biology

- Conference: International Conference on Plant Synthetic Biology and Bioengineering
- Year: 2016
- Proceeding: International Conference on Plant Synthetic Biology and Bioengineering
- Group: General Submissions
- Session:

  Rational Design and mining of natural systems for plant synthetic biology components

- Time:

  Friday, December 16, 2016 - 12:25am-12:50am

Authors:
Shih, P., Joint BioEnergy Institute, Lawrence Berkeley National Laboratory
Plant synthetic biology efforts have been hampered by a dearth of versatile transformation vectors, DNA part libraries, and efficient assembly strategies. Here, we present a strategy utilizing *in vivo* yeast homologous recombination to assemble multiple gene cassettes to facilitate plant metabolic engineering, which we have named jSTACK. In doing so, we have also generated a library of DNA parts consisting of promoters, genes, and terminators that will be publicly available as a resource to the plant synthetic biology community. We demonstrate how this method can facilitate pathway engineering of molecules of pharmaceutical interest, production of potential biofuels, and shuffling of disease resistance traits between crop species. *In vivo* homologous recombination has been leveraged for the large-scale DNA assembly of synthetic chromosomes and genomes in microbes. Likewise, our approach extends this technology to plants, providing a powerful alternative to conventional approaches for stacking traits and genes to address many impending agricultural challenges.

Design and Implementation of Genetic Circuits for Complex Biological Functions in Plants

- Conference: International Conference on Plant Synthetic Biology and Bioengineering
- Year: 2016
- Proceeding: International Conference on Plant Synthetic Biology and Bioengineering
- Group: General Submissions
- Session:

  Rational Design and mining of natural systems for plant synthetic biology components

- Time:

  Friday, December 16, 2016 - 12:50am-1:15am

Authors:
Antunes, M., Colorado State University
Medford, J. I., Colorado State University
Morey, K., Colorado State University
Schaumberg, K., Massachusetts Institute of Technology (MIT)
Plants and photosynthetic organisms provide a means towards sustainable life on Earth through the production of bio-based systems for food, materials, energy and more. Efficient and predictable plant-based production systems must rely on the implementation of complex synthetic biological circuits for gene regulation and decision-making strategies in cells and tissues, especially through use of digital and analog computation. Achieving predictable function of these circuits is further complicated by the multicellular nature of plants. Using computational models and quantitatively defined genetic parts, we have designed and implemented synthetic genetic circuits that function in whole plants with predictability. Computationally re-designed ligand-binding proteins are powerful tools for the development of biosensors for small molecules. By linking the

ligand binding event to transcriptional activation of downstream genes, we have shown that plants can be engineered with sensitive and specific biosensing capabilities. In the absence of ligand, the sensing protein is rendered unstable, and does not accumulate in the cell. Upon ligand binding, the protein is stabilized and binds specifically to operator sequences in the promoter region of genes of interest. The system is modular, in the sense that input can be provided by newly re-designed small molecule binding proteins. Reporter genes, as well as functional genes, can be used as output. This biosensing system can be used to detect hazards in the environment, as well as a controllable component of more complex genetic circuits. Biosensors in multicellular organisms may suffer from a high detection limit, where high concentrations or long exposures to the ligand are necessary for triggering a response. To circumvent this limitation, we have also engineered positive feedback genetic circuits that amplify the input signal several fold in Arabidopsis plants. These circuits may also provide â€œmemoryâ€ of detection, *i.e.*, transient exposure to the ligand results in a permanent response. The next generation of transgenic crops will contain small gene networks that have decision-making properties, with the ability to reshape patterns of plant metabolism and growth, for applications that range from security to bio-manufacturing.

Development of a Two-Component System for Signal Propagation in Plants

- Conference: International Conference on Plant Synthetic Biology and Bioengineering
- Year: 2016
- Proceeding: International Conference on Plant Synthetic Biology and Bioengineering
- Group: General Submissions
- Session:

  Rational Design and mining of natural systems for plant synthetic biology components

- Time:

  Friday, December 16, 2016 - 1:15am-1:40am

Authors:
Schreiber, T., Leibniz Institute of Plant Biochemistry
Tissier, A., Leibniz Institute of Plant Biochemistry

Our objective is to engineer orthogonal signal propagation circuits in plants. One approach to this is the development of binary (or two-component) systems based on customized transcription factors. Transcription Activator-like Effectors (TALEs) and CRISPR/Cas nucleases are proteins with programmable sequence specific DNA binding, forming the basis for orthogonal DNA binding specificity in a given host organism. The transcription-based two component systems we designed are inspired by the yeast two-hybrid assay, where a DNA binding domain and a transcription activation domain are brought together via protein-protein interaction. For signal propagation, DNA binding units based on TALEs or CRISPR/Cas can be constitutively expressed and can interact with an independently expressed activation unit. By conferring intercellular mobility to the activation unit, signal propagation can be achieved. We will present our design strategies and recent results for the engineering of binary transcription activation for signal propagation.

Determining the Principles behind Tissue-Specific Gene Expression

- Conference: International Conference on Plant Synthetic Biology and Bioengineering
- Year: 2016
- Proceeding: International Conference on Plant Synthetic Biology and Bioengineering
- Group: General Submissions
- Session:

  Rational Design and mining of natural systems for plant synthetic biology components

- Time:

  Friday, December 16, 2016 - 1:40am-2:05am

Authors:
Megraw, M., Oregon State University

In our laboratory, we have recently created a genome-wide dataset that includes both Transcription Start Sites (TSSs) and Open Chromatin (OC) regions in Arabidopsis roots, as well as machine learning models trained on these datasets that allow us to: 1) very accurately predict genomic locations that will express highly, and 2) derive from the model those cis-regulatory elements (CREs) that are most predictive of expression for each highly-expressing TSS location. The analogous dataset and modeling concept applied to leaves successfully learns a CRE-based model for leaf TSSs. What distinguishes leaf-specific from root-specific characteristics in endogenous plant promoters, and what do these characteristics reveal about synthetic design principles for tissue-specific promoters? We have taken the first step toward addressing these questions in building a model that can readily distinguish a highly expressed root-specific TSS from a leaf-specific TSS. In this talk, we will explain what we have learned by interpreting this model, and discuss the results of several initial *in vitro* tests of these concepts.

•

Rational Design and Biological Complexity: Quantitative Measurement and Prediction of Synthetic Circuits Using Plant Protoplasts

ICPSBB
International Conference on Plant Synthetic Biology and Bioengineering

- Conference: International Conference on Plant Synthetic Biology and Bioengineering
- Year: 2016
- Proceeding: International Conference on Plant Synthetic Biology and Bioengineering
- Group: General Submissions
- Session:

  Design and Predictive Function of Plant Gene Circuits

- Time:

  Friday, December 16, 2016 - 12:00am-12:25am

Authors:
Medford, J. I., Colorado State University
Schaumberg, K., Massachusetts Institute of Technology (MIT)
Antunes, M., Colorado State University
Kassaw, T., Colorado State University
Zalewski, C., Front Range Biosciences
Plant synthetic biology is of great interest for development of sustainable green technologies to meet basic needs. A key step in the rational design of synthetic networks is the quantitative characterization of components to enable predictive modeling of the synthetic network properties. However this poses special difficulties for plants since

stably transforming plants is time consuming and can take months. An alternative strategy is the use of transient protoplast assays for quantitative testing of components. We developed an experimental test bed for characterizing externally inducible repressors using protoplasts in 96-well plates, but found that transient protoplast assays show significant experimental variability that makes direct quantitative comparisons yield incorrect results. I will discuss some of the sources of variability as well as a mathematical model we developed and then used to normalize the data and make quantitative comparisons between different inducible repressors. Despite some remaining uncertainties we showed that protoplast assays could approximately predict quantitative properties of synthetic circuits in stably transformed plants. We tested hundreds of repressible promoters built by the Medford lab, and carried out a statistical analysis of the quantitative data to uncover design principles for building synthetic inducible repressors in plants (Nature Methods, v13, pp94â€"100, (2016)). I will also discuss our current attempts to use flow cytometry for protoplast experiments. The next step in rational design involves using the numbers we obtain from our test-bed to predict the properties of larger networks that are assembled from these quantitatively characterized components. In our case we were interested in finding best combinations for a genetic toggle switch in a plant. An important question that arises here is how well *in silico* predictions can be expected to match experiments in complex organisms. I will discuss statistical methods we use for making such predictions as well as some of the complexities of real biological systems that pose challenges that need to be overcome.

Computational Modeling of Cytoskeletal and Cell Wall Systems during Cell Morphogenesis

- Conference: International Conference on Plant Synthetic Biology and Bioengineering
- Year: 2016
- Proceeding: International Conference on Plant Synthetic Biology and Bioengineering
- Group: General Submissions
- Session:

  Design and Predictive Function of Plant Gene Circuits

- Time:

  Friday, December 16, 2016 - 12:25am-12:50am

Authors:
Szymanski, D.
In plants, the actin and microtubule cytoskeletons control cell morphogenesis by modulating local cell wall properties. We previously used a combination of finite element (FE) computational modeling and multivariate live cell imaging to demonstrate that the actin filament nucleator ARP2/3 generates actin meshworks in a microtubule depletion zone at the branch apex of developing trichomes (Yanagisawa et al., 2015). The actin meshwork appears to have two functions: 1) it organizes actin bundle networks for bidirectional transport and for uniform distribution of secretory organelles; 2) it dynamically links the geometry of the cell wall with the size of the microtubule depletions zone to enable branch tapering during elongation. At present the cellular mechanisms by which ARP2/3 is clustered and activated is not known. In this study, we focus on the upstream regulators of ARP2/3 in the trichome system. Rho of plant (ROP) small GTPases function as critical regulators to re-organize cytoskeletons. SPIKE1 is a guanine nucleotide exchange factor, which converts the inactive ROP-GDP form into the active ROP-GTP form. ROP-GTP activates the WAVE/SCAR regulatory complex

(W/SRC) which then unleashes the actin filament nucleation activity of ARP2/3. Here we report on the localization and dynamics of a fully functional YFP-tagged version of SPIKE1. We show that SPIKE1 and activated ROP-GTP localize at the trichome branch apex to mediate cell tapering. We also discover that SPIKE1, W/SRC and ARP2/3 cluster beneath the plasma membrane at putative ER-subdomains that can persist for tens of minutes. W/SRC localization is altered in *spike1* trichomes and pavement cells, indicating that SPIKE1 is required for normal clustering and activation of ARP2/3. Interestingly, polarized accumulation of SPIKE1 is lost in developing trichomes when microtubules are depolymerized. These results indicate that SPIKE1 works in close cooperation with microtubules to determine an ARP2/3 activation domain that dynamically regulates cell wall heterogeneity and cell morphogenesis. These results will be discussed in the context of discovering growth control modules that may be adapted to different cell types to engineer desired cell shapes.

Development and Application of Synthetic Biology Tools to Modulate Decarboxylation in Arabidopsis

ICPSBB
International Conference on Plant Synthetic Biology and Bioengineering

- Conference: International Conference on Plant Synthetic Biology and Bioengineering
- Year: 2016
- Proceeding: International Conference on Plant Synthetic Biology and Bioengineering
- Group: General Submissions
- Session:

  Design and Predictive Function of Plant Gene Circuits

- Time:

  Friday, December 16, 2016 - 12:50am-1:15am

Authors:
De Paoli, H. C., Oak Ridge National Laboratory
Yang, X., Oak Ridge National Laboratory
Tuskan, G., Oak Ridge National Laboratory
Tschaplinski, T. J., Oak Ridge National Laboratory
Hettich, R., Oak Ridge National Laboratory
Engle, N. L., Oak Ridge National Laboratory
Borland, A., Newcastle University
Abraham, P. E., Oak Ridge National Laboratory
Sun, Y., Oak Ridge National Laboratory
We recently described an advanced system, named TNT-cloning, for joining DNA fragments from a universal library that automatically keeps open reading frames (ORFs) and does not require linkers, adaptors, sequence homology, amplification or mutation of fragments containing restriction enzyme sites in order to work properly. This all-inclusive method, which is enhanced by a unique buffer formulation, increases the current capabilities of testing and sharing multi-gene circuitry assembled from various DNA

segments. Here, we present an improvement of the method for higher cloning efficiency and its expansion for use in multiple model systems. Also, we applied such genetic circuitry to create multi-gene loss-of-function lines and a proof-of-concept for manipulation of decarboxylation in *Arabidopsis*. We targeted all the four malic enzymes (NADP-ME1-4) in *Arabidopsis*, which are responsible for malate breakdown into pyruvate and CO_2 in plant cells. Diel regulation of the malic enzymes, and therefore the decarboxylation step, are key steps acquired by evolution for the adaptation of CAM (Crassulacean Acid Metabolism) plants to water-limited environments. Therefore, the capability of manipulating diel carboxylation/decarboxylation in plants is critical for CAM-into-C_3 engineering to increase water use efficiency (WUE) and drought tolerance in C_3 plants. NADP-ME loss-of-function lines showed increased photosynthetic rate, higher net CO_2 fixation rates, higher biomass yield and improved WUE with higher drought tolerance relative to the wild-type *Arabidopsis* plants. Furthermore, diel metabolomics analysis showed the accumulation of malate is temporal and becomes readily converted into sugars/carbohydrates along with precursors of cell wall biogenesis. To demonstrate the feasibility of creating an inducible and reprogrammable metabolic state for malate decarboxylation, we complemented the NADP-ME loss-of-function lines with a dark-induced promoter and monitored gas exchange during a 24-hour period. Our results demonstrate the robustness of a controller, comprised of YES (1), AND (2) and XOR (2) gates, to effectively link 3 inputs, which includes an ON-OFF switch, and generate two distinct outputs for diel control of intracellular CO_2 in leaves. Taken together, our research leveraged synthetic biology tools for bioengineering of complex traits using genetic circuitry, particularly in plants.

Reprogramming the plant form: Using synthetic signalling to study and re-engineer development

- Conference: International Conference on Plant Synthetic Biology and Bioengineering
- Year: 2016
- Proceeding: International Conference on Plant Synthetic Biology and Bioengineering
- Group: General Submissions
- Session:

  Design and Predictive Function of Plant Gene Circuits

- Time:

  Friday, December 16, 2016 - 1:15am-1:40am

Authors:
Khakhar, A., Johns Hopkins University
Lemmex, A.
Klavins, E., University of Washington
Nemhauser, J., University of Washington
Human society is dependent on plant products for essentials such as food, textiles, lumber and pharmaceuticals. However, due to a rapidly changing agricultural environment and a steadily increasing demand for resources, we face a future where access to these essentials may be severely curtailed. One strategy to address these challenges is to develop tools that allow us to rationally re-engineer plants to be more productive and robust. In this work, we present a set of strategies to re-engineer development using synthetic signaling systems to alter how plants interpret and respond to hormonal cues, specifically of the plant hormone auxin, which plays a central role in development. We demonstrate how synthetic auxin-sensitive promoters can be used in conjunction with mathematical modeling to rationally alter a range of developmental processes, such as phyllotaxy, in *Arabidopsis thaliana*. To address the weaknesses of this approach, namely

the lack of tissue specificity and fine grain tunability of auxin sensitivity, we developed a novel class of auxin-degradable CRISPR-based transcription factors (ADCTFs). We demonstrate that our modular ADCTF platform has the potential to tunable regulate genes based on endogenous auxin fluxes. It could also be used to visualize the fluxes of auxin during development, in ways that complement existing reporters, to further elucidate the role of these fluxes in development. Future applications of this technology include the rapid domestication of wild plants by making them less bushy and making plants with programmable root architectures for drought resistance.

Synthetic Programming of Combinatorial Gene Regulation in Plant Development

- Conference: International Conference on Plant Synthetic Biology and Bioengineering
- Year: 2016
- Proceeding: International Conference on Plant Synthetic Biology and Bioengineering
- Group: General Submissions
- Session:

  Design and Predictive Function of Plant Gene Circuits

- Time:

  Friday, December 16, 2016 - 1:40am-2:05am

Authors:
Pierre-Jerome, E.
Benfey, P.
Specification and maintenance of cell identity relies on the complex interplay of transcription factors to generate diverse spatiotemporal patterns of gene expression. In Arabidopsis, the transcription factors SHORTROOT (SHR) and SCARECROW (SCR) are required for root radial patterning. Together they directly regulate CYCD6, a cell-cycle regulator, to orchestrate the formation of the ground tissue. Since SHR activates both SCR and CYCD6, the SHR-SCR-CYCD6 circuit forms a feed-forward loop (FFL), one of the most prevalent motifs in gene regulatory networks. Despite the overrepresentation of feed-forward loops in developmental networks, current understanding of FFL behavior is largely theoretical. To directly test feed-forward loop function in a developing plant, I am rewiring the native CYCD6 circuit and building a synthetic circuit to regulate CYCD6 in a feed-forward manner.

Evolvability and Functional Divergence in the Auxin-Signaling F-Box Family

- Conference: International Conference on Plant Synthetic Biology and Bioengineering
- Year: 2016
- Proceeding: International Conference on Plant Synthetic Biology and Bioengineering
- Group: General Submissions
- Session:

  Design and Predictive Function of Plant Gene Circuits

- Time:

  Friday, December 16, 2016 - 2:05am-2:30am

Authors:
Wright, R. C., University of Washington
Zahler, M. L.
Gerben, S. R., University of Washington
Nemhauser, J., University of Washington
Auxin is a small molecule hormone that shapes plant development and homeostasis. The evolution of more complex body plans in land plant lineages has been paralleled by gene duplication and divergence within nuclear auxin-signaling networks. Mapping how the structure/function relationship of these signaling proteins affects plant phenotype would allow rational engineering of plant growth, development and reproduction. Here, we determined this mapping for a subset of permissive mutations found in the auxin receptor F-box family of wild accessions of the reference plant *Arabidopsis thaliana*. To do this, we quantified the functional variation between the natural F-box variants and the phenotypic variation in accessions expressing these variants. In several cases quantitative changes in auxin sensitivity at the whole plant level were predicted by synthetic functional assays in yeast. Our findings revealed a strong connection between sequence

variation in paralogs, altered molecular performance and phenotypic diversity at the organism-level. This approach provides a new method for harnessing the wealth of sequence information to guide future engineering efforts of diverse signaling pathways which rely on ubiquitin-mediated protein turnover.

Production of Halogenated Indigo Precursors in Metabolically Engineered Tobacco Plants

- Conference: International Conference on Plant Synthetic Biology and Bioengineering
- Year: 2016
- Proceeding: International Conference on Plant Synthetic Biology and Bioengineering
- Group: General Submissions
- Session:

  Plant foundry: Biopolymers, Biopharmaceuticals and Medically Effective Secondary Metabolites (Vaccines, antibodies, and metabolites)

- Time:

  Friday, December 16, 2016 - 12:00am-12:25am

Authors:
Warzecha, H.
Fräbel, S.

Indigogenic glycosides are useful chromogenic substrates for histochemistry and molecular biology. Most prominent is X-Gal (5-Bromo-4-chloroindoxyl-ß-D-galactoside) used for blue white selection during molecular cloning procedures in bacteria. Although the precursor indican is a natural product widespread in a large number of indigo plants, the halogenated derivatives mainly utilized as enzyme substrates needs to be chemically synthesized. Synthesis of halogenated aromatic components require numerous toxic precursors and intermediates as well as organic solvents, rendering the whole process potentially harmful to the environment. We established a system for the *in-planta* biosynthesis of a large array of indigogenic glucosides. Engineering of a three different bacterial halogenases along with tryptophanase and a monooxygenase lead to the formation of specifically chlorinated indican derivatives. Moreover, combination of halogenases enabled chlorination at multiple sites and co-administration of bromine enabled the formation of brominated indican derivatives. Also different subcellular compartments for optimal precursor supply as well as product formation were evaluated. Taken together, plants provide a suitable platform for the production of halogenated fine chemicals.

•

Red Biotechnology Potential of Taxol-Producing Endophytic Fungus Lasodiplodia Theobromea Isolated from a Non-Taxol-Producing Plant

ICPSBB
International Conference on Plant Synthetic Biology and Bioengineering

- Conference: International Conference on Plant Synthetic Biology and Bioengineering
- Year: 2016
- Proceeding: International Conference on Plant Synthetic Biology and Bioengineering
- Group: General Submissions
- Session:

  Plant foundry: Biopolymers, Biopharmaceuticals and Medically Effective Secondary Metabolites (Vaccines, antibodies, and metabolites)

- Time:

  Friday, December 16, 2016 - 12:25am-12:50am

Authors:
Sah, B., Indian Institute of Sceince
S., K.
C., J.
Taxol®(generic name Paclitaxel) a plant-derived anticancer drug, widely used against breast,ovarian and lung cancer,was originally isolated from the Pacific Yew,*Taxus brevifolia*.The limited supply of the drug prompted research to find alternative sources,such as chemical synthesis,tissue and cell culture of the *Taxus* species both of which are expensive and yield low levels.Thus,a sustainable, economical and unconventional alternative source of Taxol such as endophytic fungi has been actively researched.

A rich variety of endophytic fungi from Taxol- producing and non-Taxol-producing plants are known to produce Taxol in axenic cultures. *Lasodiplodia theobromea* species are among the most important endophytic fungi with biotechnological potential. In search of Taxol-producing endophyitc fungi from non-Taxol producing plant such as *Piper nigrum*, *Zizypus sp* and *Mesura ferrae* we isolated total of eight *L.theobromea,* three of which were found to produce Taxol.One of these three *L.theobromea* isolated from non-Taxol-producing *Piper nigrum* was selected in this study. Its identity was confirmed by morphological characteristics and ITS rDNA sequence analyses. Taxol produced by *L.theobromea* was shown to be identical to the authentic Taxol as analysed by TLC, HPLC, LC-MS and MS/MS analyses. The quantity of Taxol produced by the fungus was estimated to be 247μg/L.

As an evidence to support the independent production of Taxol by the *L.theobromea*,the gene encoding the 10-deacetylbaccatinIII-10-0-acetyltransferasse(DBAT), which catalyzing the penultimate acetylation in the Taxol biosynthesis pathway, was cloned and sequenced from *L.theobromea.* Additionally the same gene was cloned from the Taxol- producing *Taxus celebica* plant. The predicted amino sequence of the *L.theobromea* and the plant DBAT shows high homology to DBATs fromTaxol- producing *Taxus celebica and Taxus cuspidate* plants as well as the endophytic fungus *Cladosporium.* Further, the ORFs of *L.theobromea* and *T.celebica* DBAT were heterologously expressed in *E.coli* and the purified recombinant DBAT enzymes were shown to be capable of converting 10-deacetylbaccatinIII to baccatinIII. Therefore, Taxol biosynthetic DBAT gene in *L.theobromea* seems to have evolved independently and thus rule out horizontal gene transfer between the host plant and its endophytes.

This study can provide a road map to improve the production level and minimise production cost of Taxol at industrial level by implementing techniques of synthetic biology.

Development of a High-Throughput Physcomitrella patens Transformation System for Biosynthetic Applications

- Conference: International Conference on Plant Synthetic Biology and Bioengineering
- Year: 2016
- Proceeding: International Conference on Plant Synthetic Biology and Bioengineering
- Group: General Submissions
- Session:

  Plant foundry: Biopolymers, Biopharmaceuticals and Medically Effective Secondary Metabolites (Vaccines, antibodies, and metabolites)

- Time:

  Friday, December 16, 2016 - 12:50am-1:15am

Authors:
Anderson-Furgeson, J., Taxa Biotechnologies
Evans, A.
Moon, J.
Kain, J., Taxa Biotechnologies
A key challenge for plant synthetic biology is the development of a platform for efficient multi-gene engineering. The moss Physcomitrella patens already has several desirable features for a biosynthetic platform, such as precise genome engineering via homologous recombination, easy upscaling of homogenous cultures, and a fully sequenced genome. However, a lack of methods for rapid and easy multi-gene engineering keeps this system from realizing its potential. Here we seek to fix two major limitations: 1) the lack of a

genetic regulatory element toolkit and 2) a transformation process that is bottlenecked by labor. Creating a transgenic moss strain expressing more than one gene product necessitates the use of several promoters and terminators to prevent inappropriate recombination between multiple copies. Thus, new promoters and terminators need to be screened and catalogued. Automation of moss transformation and tissue culture techniques will not only aid our genetic element screen, but will accelerate the advent of a useful and productive moss synthetic biology platform.

A Synthetic Biology Yeast Platform for the Biosynthesis of Plant Phenolic Diterpenes

ICPSBB
International Conference on Plant Synthetic Biology and Bioengineering

- Conference: International Conference on Plant Synthetic Biology and Bioengineering
- Year: 2016
- Proceeding: International Conference on Plant Synthetic Biology and Bioengineering
- Group: General Submissions
- Session:

  Plant foundry: Biopolymers, Biopharmaceuticals and Medically Effective Secondary Metabolites (Vaccines, antibodies, and metabolites)

- Time:

  Friday, December 16, 2016 - 1:15am-1:40am

Authors:
Tissier, A., Leibniz Institute of Plant Biochemistry
Scheler, U.
The phenolic diterpenes carnosic acid (CA) and carnosol (CO) are of special interest due to their use as food additive and various biological activities. They are present in species of the Lamiaceae family such as rosemary (*Rosmarinus officinalis*) and various species of sage (e.g. *Salvia fruticosa*). Several studies over the past years report on the anti-oxidative, anti-cancer, anti-microbial and anti-inflammatory activities of PDs. Furthermore, the biosynthesis of the related tanshinone diterpenoids from *Salvia miltiorrhiza*, which also have high antioxidant, antibiotic and anti-inflammatory activities, is presumed to proceed via CA. Thus, elucidation of the biosynthesis of CA represents a key step in understanding the genesis of phenolic diterpenes and will provide the basis for metabolic engineering of this class of compounds. To facilitate the co-expression of multiple genes in the yeast *Saccharomyces cerevisiae*, a Golden Gate based

modular cloning system was established with a set of parts including promoters and terminators.

Using various EST databases from *Rosmarinus officinalis* and *Salvia fruticosa* we have isolated cytochrome P450 genes, *hydroxyferruginol synthases* (*CYP76AH22-24*) and *C20-oxidases* (*CYP76AK6-8*), which were tested for their activity in yeast. Co-expression of the genes (*GGPPS*, *CPS*, *MS*) for the biosynthesis of miltiradiene, the diterpene precursor of CA, along with the *CYP76AH22-24* genes leads to the production of the pathway intermediates ferruginol and 11-hydroxyferruginol. These are natural products of rosemary, and were identified by mass spectrometry and NMR. Using modeling-based mutagenesis by comparing the CYP76AH22-24 enzymes to CYP76AH1, a ferruginol synthase from *S. miltiorrhiza* that cannot produce hydroxyferruginol, we identified three amino acid residues which are required and sufficient for the C11 hydroxylation of ferruginol. Lastly, co-expression of the C20-oxidases leads to the complete conversion of 11-hydroxyferruginol to CA (Scheler et al., 2016). Additional enzymes are being expressed to expand the repertoire of modifications on the phenolic diterpene skeleton, thereby constituting a platform for combinatorial biosynthesis of this class of biologically active diterpenoids.

Reference: Scheler et al., (2016), Elucidation of the biosynthesis of carnosic acid and its reconstitution in yeast, *Nature Communications* (in press).

Cutting-Edge Methods and Tools for Plant Genome Editing

- Conference: International Conference on Plant Synthetic Biology and Bioengineering
- Year: 2016
- Proceeding: International Conference on Plant Synthetic Biology and Bioengineering
- Group: General Submissions
- Session:

  Plant genome editing (CRISPR systems)

- Time:

  Saturday, December 17, 2016 - 12:00am-12:25am

Authors:
Cermak, T.
Curtin, S.
Gil-Humanes, J.
Cegan, R.
Starker, C.
Kono, T.
Konecna, E.
Mathre, J.
Greenstein, B.
Stupar, R.
Voytas, D.

The development of DNA modifying enzymes with programmable DNA-binding specificities, derived from transcription activator-like effectors (TALEs) and the CRISPR/Cas9 system have opened new opportunities for more effective engineering of plant genomes. Using TALENs and CRISPR/Cas9 nucleases, gene knockouts can now be efficiently created in any plant species amenable to transformation. However, to fully exploit the potential of the gene editing technologies for synthetic biology, more advanced tools are needed to induce more complex modifications, such as gene replacement and multiplex gene editing, and to control transcriptional regulation. Moreover, parallel use of different gene editing techniques might be necessary to build and program genetic circuits and to make it possible to engineer whole genetic pathways. Here we introduce a comprehensive suite of methods and reagents that utilize both TALE(N)s and CRISPR/Cas9 for genome engineering in both monocot and dicot plant systems. The reagents are organized into an easy-to-use modular cloning system that offers the flexibility of combining different functional modules for different applications within a single construct. We have used this system to efficiently induce targeted gene insertions, to simultaneously edit multiple genes, and to create large deletions in a variety of plant species including tomato, wheat, barley, and cassava. Examples of successful genome engineering efforts will be presented.

Crispr-Cas9 or Talen Mediated Genome Editing in Sugarcane

- Conference: International Conference on Plant Synthetic Biology and Bioengineering
- Year: 2016
- Proceeding: International Conference on Plant Synthetic Biology and Bioengineering
- Group: General Submissions
- Session:

  Plant genome editing (CRISPR systems)

- Time:

  Saturday, December 17, 2016 - 12:25am-12:50am

Authors:
Altpeter, F., University of Florida - IFAS
Oz, T., University of Florida - IFAS
Kannan, B., University of Florida - IFAS
Karan, R., University of Florida - IFAS
Jung, J. H., University of Florida - IFAS
Merotto, A., University of Florida - IFAS
Sugarcaneâ€™s exceptional photosynthetic efficiency results in superior biomass and sugar production in support of the emerging bio-economy. While designer nucleases like CRISPR-Cas9 and TALEN can now be widely used to precisely target a specific DNA sequence and cause cleavage in vivo, there is no control over the subsequent NHEJ process that takes place. Hence, although the site of mutation is highly specific, the resultant structure of the mutated locus is largely random and consists of indels of unspecified size and sequence.Â Associated frameshift mutations will result in loss of function and cause desirable phenotypes in sugarcane if a sufficient number of alleles/copies in the highly-polyploid genome is co-edited.

However, in contrast to these â€œloss of function mutantsâ€ many synthetic biology applications will require the precise conversion of inferior to superior alleles to gain function. Truly precise sequence engineering of an endogenous locus requires template mediated homologous recombination (HR) which competes with the more efficient but error prone NHEJ during DNA repair. We will describe alternative approaches to genome editing in sugarcane, as well as faithful transmission of genome modifications to vegetative progenies and their field performance.

Nanoparticles As Biomolecular Cargo Transporters in Plants and Plastids

- Conference: International Conference on Plant Synthetic Biology and Bioengineering
- Year: 2016
- Proceeding: International Conference on Plant Synthetic Biology and Bioengineering
- Group: General Submissions
- Session:

  Plant genome editing (CRISPR systems)

- Time:

  Saturday, December 17, 2016 - 12:50am-1:15am

Authors:
Demirer, G. S., Koc University
Landry, M., University of California Berkeley
The field of targeted plant genome editing has experienced exciting progress since the production of the first transgenic plants. Recent advances in targeted genome engineeringâ€“ development of zinc finger nucleases (ZFNs), TAL effector nucleases (TALENs), and CRISPR/Cas systemsâ€“ allow modification of DNA sequences of interest, once inside living cells.[1] To introduce genome engineering tools into plant cells and plastids, several methods have been established, such as Agrobacterium or polyethylene glycol (PEG) mediated gene delivery, electroporation, biolistic particle gun, and microinjection. However, current methods are limited by host range limitations, low transfection efficiency, toxicity, and in most cases, transfection limited to protoplast cultures where the cell wall is removed, and not to the mature whole plant.[2] We explore the use of high-aspect ratio nanomaterials for efficient through-cell-wall gene transfer into all desired genotypes of any plant species.

Nanomaterials, with distinct and tunable physical properties (size, shape, and surface charge), effectively deliver biological cargoes to traditionally hard-to-access intracellular locations, such as walled plant cells. We have previously shown that certain nanomaterial formulations can penetrate the mature plant cell, and also the chloroplast.[3,4] Here, we present recent results on the passive delivery of nanoparticles, loaded with plasmid DNA, into walled plant cells and protoplasts. In particular, we examine the delivery mediated by single-walled carbon nanotubes (SWCNT) functionalized with GFP encoding plasmids, and the resulting GFP expression in the transfected cells of mature arugula plants. Our work provides a promising tool for generic, targeted, and passive transport of biomolecules (genes, proteins, etc.) into mature plant cells for targeted genome editing and plastid engineering.

References

1. Li, J.-F. *et al.* Multiplex and homologous recombination-mediated genome editing in Arabidopsis and Nicotiana benthamiana using guide RNA and Cas9. *Nature biotechnology* **31**, 688-691 (2013).

2. Potrykus, I. Gene Transfer to Plants: Assessment of Published Approaches and Results. *Annual Review of Plant Physiology and Plant Molecular Biology* **42**, 205-225, doi:10.1146/annurev.pp.42.060191.001225 (1991).

3. Giraldo, J. P. *et al.* Plant nanobionics approach to augment photosynthesis and biochemical sensing. *Nat Mater* **13**, 400-408, doi:10.1038/nmat3890

4. Wong, M. H. *et al.* Lipid Exchange Envelope Penetration (LEEP) of Nanoparticles for Plant Engineering: A Universal Localization Mechanism. *Nano Letters* **16**, 1161-1172, doi:10.1021/acs.nanolett.5b04467 (2016).

Functional Analyses of Cotton (Gossypium hirsutum L.) Immature Fiber (im) Mutant Whose Phenotype Grown Under Normal Conditions Is Similar to Wild Type Cotton Grown Under Severe Stress

- Conference: International Conference on Plant Synthetic Biology and Bioengineering
- Year: 2016
- Proceeding: International Conference on Plant Synthetic Biology and Bioengineering
- Group: General Submissions
- Session:

  Agricultural Sense and Adaptation Systems for Stress Response

- Time:

  Saturday, December 17, 2016 - 12:25am-12:50am

Authors:
Kim, H. J.
Delhom, C., USDA-ARS-SRRC
Liu, Y., USDA-ARS-SRRC
Rodgers, J., USDA-ARS-SRRC
Fang, D.
Kim, S. H., Pennsylvania State University
Cotton fiber wall thickness is an important property of determining values of cotton fibers. Cotton fiber wall is mainly composed of secondary cell wall consisting of almost pure cellulose. How fiber cell wall development affects the degree of fiber wall thickness

is not well understood. An immature fiber (*im*) mutant and its near isogenic wild type, Texas Marker-1 (TM-1) differing in fiber wall thickness provide a unique way of studying structural properties of cellulose. The wild type TM-1 produces thick and mature fiber walls under normal environmental conditions, but generated thin and immature fiber walls under severe stress conditions. The *im* mutant produced thin and immature fiber walls under normal conditions. Transcriptome analysis also showed that the differentially expressed genes (DEGs) in the *im* mutant fibers grown under normal stress conditions were similar to those in wild type cotton fibers grown under severe stress conditions. The low degree of fiber cell wall thickness of the *im* mutant fibers is associated with dysregulation of the genes involved in stress responses and cellular respiration. Phenotyping analyses showed common levels in cellulose content, crystallinity index, crystal size, matrix polymer composition, fiber perimeter, or width between the isogenic fibers, whereas there were significant differences in their linear density and the lateral order between their cellulose microfibrils (CMFs). The results provide an insight of understanding how cellulose biosynthesis process can be affected by environments during cotton fiber development.

Insight into the Plant Stress Adaptation Strategies during Tripartite Interaction of Plant-Pathogen and Nanoparticles Using Comparative Proteomics Approach

ICPSBB
International Conference on Plant Synthetic Biology and Bioengineering

- Conference: International Conference on Plant Synthetic Biology and Bioengineering
- Year: 2016
- Proceeding: International Conference on Plant Synthetic Biology and Bioengineering
- Group: General Submissions
- Session:

  Agricultural Sense and Adaptation Systems for Stress Response

- Time:

  Saturday, December 17, 2016 - 12:50am-1:15am

Authors:
Kumari, M.
Agarwal, L., CSIR-NBRI
Pandey, S., Csir Nbri
Mishra, A.
Nautiyal, C.
Biosynthesized silver nanoparticles have emerged as an economical approach to combat pathogenic microbes but still remain in nascent stage in field of agriculture. The plants responses during tripartite interaction of plant pathogen and nanoparticles also remain largely unknown. This study demonstrates the use of biosynthesized silver nanoparticles (Spherical, 2-5nm) in combating black spot disease in model plant pathosystem of *Arabidopsis thaliana* and *Alternaria brassicicola*. The particles reduced disease severity

by 70-80% at a low concentration of 5μg/ml without showing pytotoxicity. It also elicited plant immunity as evident by significant reduction in reactive oxygen species, lipid peroxidation, proline content and consequently decrease in stress enzymes viz. superoxide dismutase, catalase, guaiacol peroxidase, polyphenol oxidase and phenylalanine ammonia lyase. Further, mass spectrometry based comparative proteomics approach was carried out to gain an insight into the protein dynamics during the interaction. After two dimensional gel electrophoresis, a total of 599 proteins expressed during the interaction, in which 144 were found to be differential. 117 differential proteins were identified with significant scores. Out of 117 significant proteins with differential expression, 44% were involved in bioenergy and metabolism (BEM), 20% were in plant defence related proteins, 14, 10 and 12% were involved in cell signalling, storage and biogenesis, and miscellaneous functions respectively. It was observed that expression level of protein involved in protein biogenesis were enhanced to produce abundant antimicrobial metabolites and proteins involved in early stage of plant defence so that production of ROS is minimized, and thus the proteins of late defence were downregulated. Biosynthesized silver nanoparticles performed dual functions to combat the pathogen attack by killing plant pathogen and eliciting plant immunity by altering plant proteome.

Fruit Crop Engineering with Igt Family Genes

- Conference: International Conference on Plant Synthetic Biology and Bioengineering
- Year: 2016
- Proceeding: International Conference on Plant Synthetic Biology and Bioengineering
- Group: General Submissions
- Session:

  Agricultural Sense and Adaptation Systems for Stress Response

- Time:

  Saturday, December 17, 2016 - 1:15am-1:40am

Authors:
Guseman, J. M.
Webb, K.
Hollender, C. A.
Raines, D.
Srinivasan, C.
Dardick, C.
Plant architecture influences many aspects of plant interaction with the environment. Thus, rational design of tree shape has major implications for genetic improvement of crop species. Branch orientation, or angle, is an understudied aspect of plant architecture that affects sunlight capture in shoots, and water uptake in the roots. Here we demonstrate the role of the IGT gene family in determining branch orientation in both shoots and roots in both Arabidopsis and dicotyledonous fruit crops. Two members of the IGT family, *TILLER ANGLE CONTROL 1* (*TAC1*) and *LAZY1*, act antagonistically in the shoot to control lateral organ orientation. Loss of *TAC1* results in strongly vertical shoot growth, seen in rice tillers, maize leaves, and Arabidopsis and *Prunus* lateral buds and branches. In contrast, loss of *LAZY1* results in horizontal to prostrate growth. In the root, a third IGT gene, *DEEPER ROOTING 1* (*DRO1*), controls root angles. Rice with a truncated

DRO1 gene exhibit shallow angles across the root system. Our recent work shows that loss of *DRO1* in Arabidopsis results in wide lateral root angles, consistent with the roles of *TAC1* and *LAZY1* in dicots versus monocot shoot organs. Overexpression of *DRO1* in *Prunus domestica* (plum) results in longer root systems and the ability to root on shoot-growth culture media. IGT gene manipulation may provide important tools for bioengineering of desired plant shapes, above and below ground.

•

Engineering for Robust Photorespiratory Bypass to Improve Photosynthesis and Crop Production Under Field Conditions

- Conference: International Conference on Plant Synthetic Biology and Bioengineering
- Year: 2016
- Proceeding: International Conference on Plant Synthetic Biology and Bioengineering
- Group: General Submissions
- Session:

  Sustainable Environmental Systems and Photosynthetic Improvements

- Time:

  Saturday, December 17, 2016 - 12:00am-12:25am

Authors:
South, P. F., University of Illinois
Cavanagh, A. P., University of Illinois
Ort, D., USDA-ARS

Worldwide nearly 1 billion people are affected by hunger every day. As climate changes globally and human population increases, traditional methods of crop improvement have become less effective in adapting and improving agricultural production. In C3 crops such as wheat and rice approximately 25% of the fixed carbon dioxide is lost to photorespiration. Photorespiration is an energy expensive metabolic pathway that recycles toxic compounds produced by RubisCO oxygenation reactions. Reducing photorespiratory yield losses by 5% (i.e., to 31% for soybean and 15% for wheat) would be worth ~$500 million annually in the United States. Although photorespiration is tied to other important metabolic functions, the benefit of improving its efficiency appears to outweigh any potential secondary disadvantages. Synthetic biology has provided new opportunities in altering photorespiratory metabolism to improve photosynthetic efficiency. Indeed metabolic bypasses to photorespiration have been generated and have demonstrated improvements in growth. Using a synthetic biology approach we have assembled a series of multigene constructs that contain alternate metabolic pathways to bypass photorespiration. In addition, we designed a screen based approach to test a range of standardized parts (promoters, terminators) in the model plant *Nicotiana tabacum.* We have successfully transformed in large multigene constructs and have demonstrated detectable gene expression and improvements in biomass in replicated greenhouse and field trails. Determining robust photorespiratory bypass constructs can provide insight into next generation crops and our utilization of standard parts provide a new tool kit for plant synthetic biology to engineer improvements in photosynthetic efficiency.

Redesigning Photosynthesis to Sustainably Meet Global Food and Demand of Bioenergy

ICPSBB
International Conference on Plant Synthetic Biology and Bioengineering

- Conference: International Conference on Plant Synthetic Biology and Bioengineering
- Year: 2016
- Proceeding: International Conference on Plant Synthetic Biology and Bioengineering
- Group: General Submissions
- Session:

  Sustainable Environmental Systems and Photosynthetic Improvements

- Time:

  Saturday, December 17, 2016 - 12:25am-12:50am

Authors:
Khan, B. A., Dawood University of Engineering and Technology, Karachi
Redesigning photosynthesis to sustainably meet global food and bioenergy demand

Â

Abstract

The worldâ€™s crop productivity is stagnating whereas population growth, rising affluence, and mandates for biofuels put increasing demands on agriculture. Meanwhile, demand for increasing cropland competes with equally crucial global sustainability and environmental protection needs. Addressing this looming agricultural crisis will be one of our greatest scientific challenges in the coming decades, and success will require substantial improvements at many levels. We assert that increasing the efficiency and productivity of photosynthesis in crop plants will be essential if this grand challenge is to be met. Here, we explore an array of prospective redesigns of plant systems at various scales, all aimed at increasing crop yields through improved photosynthetic efficiency and performance. Prospects range from straightforward alterations, already supported by preliminary evidence of feasibility, to substantial redesigns that are currently only conceptual, but that may be enabled by new developments in synthetic biology. Although some proposed redesigns are certain to face obstacles that will require alternate routes, the efforts should lead to new discoveries and technical advances with important impacts on the global problem of crop productivity and bioenergy production.

The success of this process relies on its capability to exploit the almost inexhaustible reservoirs of sunlight, water, and carbon dioxide to transform photonic energy into chemical energy such as stored in adenosine triphosphate. Oxygenic photosynthesis is responsible for most of the oxygen, fossil fuels, and biomass on our planet. So, even after a few billion years of evolution, this process unceasingly supports life on earth, and probably soon also in outer-space, and inspires the development of enabling technologies for a sustainable global economy and ecosystem. The following review covers some of the major milestones reached in photosynthesis research, each reflecting lasting routes of innovation in agriculture, environmental protection, and clean energy production

Photosynthetic Antenna Engineering to Improve Crop Yields

- Conference: International Conference on Plant Synthetic Biology and Bioengineering
- Year: 2016
- Proceeding: International Conference on Plant Synthetic Biology and Bioengineering
- Group: General Submissions
- Session:

  Sustainable Environmental Systems and Photosynthetic Improvements

- Time:

  Saturday, December 17, 2016 - 12:50am-1:15am

Authors:
Melis, A.

A genetic tendency of photosynthetic organisms to assemble large arrays of light-absorbing chlorophyll antenna molecules in their photosynthetic apparatus is a survival strategy and a competitive advantage in the wild, where light is often limiting. Competition and survival in the wild requires capturing more light for self, even if in excess of what is needed to saturate photosynthesis and eventually wasted, and preventing light capture by competing neighbors. This competitive survival strategy is detrimental to production-type monocultures under bright sunlight, where over-absorption of photons by the upper canopy, and wasteful dissipation of the excess absorbed energy, are undesirable. Under bright sunlight, more than 80% of absorbed photons can thus be wasted, reducing photon use efficiencies and photosynthetic productivity to undesirably low levels. The work examines the theoretical maximum of solar energy conversion efficiency and productivity in photosynthetic systems. This will be contrasted with actual measurements of substantially lower efficiencies in a variety of photosynthetic organisms, including crop plants. Engineering a **T**runcated **L**ight-harvesting **A**ntenna size (**TLA**) in crop plants, where the number of chlorophyll molecules in the photosynthetic apparatus is genetically limited helps to alleviate excess absorption of sunlight and the ensuing wasteful dissipation of excitation energy, and to maximize solar-to-product energy conversion efficiency and photosynthetic productivity in high-density canopies of crop plants. A two-year study with wild type and TLA-tobacco as the model plant has shown TLA-tobacco biomass productivity greater by about 25% over that measured with the wild type under the same climatic and agronomic growth conditions. Thus, application of the **TLA** concept in the commercial exploitation of crop plants promises to improve growth and biomass accumulation, positively impacting crop yield, as well as plant-derived fuel, synthetic chemistry, and pharmaceutical feedstock production.

Agave and Kalanchoe Omics for Synthetic Biology to Enhance Water Use Efficiency and Drought Tolerance in Plants

ICPSBB
International Conference on Plant Synthetic Biology and Bioengineering

- Conference: International Conference on Plant Synthetic Biology and Bioengineering
- Year: 2016
- Proceeding: International Conference on Plant Synthetic Biology and Bioengineering
- Group: General Submissions
- Session:

  Sustainable Environmental Systems and Photosynthetic Improvements

- Time:

  Saturday, December 17, 2016 - 1:15am-1:40am

Authors:
Yang, X., Oak Ridge National Laboratory
Hu, R., Oak Ridge National Laboratory
Yin, H.
Liu, D., Oak Ridge National Laboratory
Schmutz, J., HudsonAlpha Institute for Biotechnology
Jenkins, J.
Shu, S.
Goodstein, D.
Abraham, P. E., Oak Ridge National Laboratory
Hettich, R., Oak Ridge National Laboratory
Cushman, J. C., University of Nevada Reno
Borland, A., Newcastle University
Tschaplinski, T. J., Oak Ridge National Laboratory

Tuskan, G., Oak Ridge National Laboratory
The ongoing and predicted climate change will likely increase drought stress with negative impact on agricultural and bioenergy crop production. Crassulacean acid metabolism (CAM) is a specialized type of photosynthesis that evolved as an adaptation to water-limited environments. CAM plants have much higher water-use efficiency (WUE) than C_3 or C_4 plants due to an inverted day/night pattern of stomatal closure/opening in CAM species relative to C_3 and C_4 species. Engineering of CAM pathway into C_3 crops using synthetic biology approach would have great potential for genetic improvement of WUE and drought tolerance for sustainable production of food and bioenergy in arid, semi-arid and marginal lands. To facilitate CAM-engineering, we generated rich omics resources (e.g., genomics, transcriptomics, proteomics, metabolomics) for *Agave* spp. and *Kalanchoe* spp., which are obligate CAM plants in the eudicotyledonous (eudicot) and monocotyledonous (monocot) lineages, respectively. Our integrative analysis of these omics data provides new insight into the evolution of CAM plants and useful knowledge to inform CAM-into-C3 engineering.

Intracellular Spectral Recompositioning of Light Increases Photosynthetic Efficiency and Reduces Light Associated Stress in a Versatile Model Diatom

ICPSBB
International Conference on Plant Synthetic Biology and Bioengineering

- Conference: International Conference on Plant Synthetic Biology and Bioengineering
- Year: 2016
- Proceeding: International Conference on Plant Synthetic Biology and Bioengineering
- Group: General Submissions
- Session:

  Sustainable Environmental Systems and Photosynthetic Improvements

- Time:

  Saturday, December 17, 2016 - 1:40am-2:05am

Authors:
Fu, W., New York University Abu Dhabi
Chaiboonchoe, A., Division of Science and Math, and Center for Genomics and Systems Biology (CGSB), New York University Abu Dhabi, Abu Dhabi, United Arab Emirates
Khraiwesh, B., New York University Abu Dhabi
Sultana, M., New York University Abu Dhabi
Jaiswal, A., Division of Science and Math, and Center for Genomics and Systems Biology (CGSB), New York University Abu Dhabi
Jijakli, K., Division of Science and Math, and Center for Genomics and Systems Biology (CGSB), New York University Abu Dhabi
Nelson, D., New York University Abu Dhabi
Salehi-Ashtiani, K., New York University Abu Dhabi

Diatoms as photosynthetic cell factories can provide the means to reach a sustainable production of petrochemical and bioactive compounds; however, a requisite to achieve this goal is to increase the species photosynthetic efficiency, which generally remains less than five percent. We have developed and implemented a strategy, herein referred to as Intracellular Spectral Recompositioning of light (or *ISR*), which through absorption of excess blue light and its intracellular emission in green spectral band, may improve light utilization. We demonstrate that *ISR* can be employed chemogenically, through use of lipophilic fluorophores, or biogenically, through expression of an enhanced green fluorescent protein (eGFP) in the model diatom *Phaeodactylum tricornutum.* Engineered *P. tricornutum* cells expressing eGFP achieved twenty-eight percent higher efficiency in photosynthesis than the parental strain under mixed red and blue light condition. Further, pond simulator experiments demonstrated that eGFP transformants could outperform their wild type parental strain by fifty percent in biomass production rate under high intensity white LED illumination. Transcriptome analysis of the engineered strain identified up-regulation of genes, such as the Light Harvesting Complex Protein X (LHCX) genes including *LHCX1*, *LHCX3*, and *LHCX4*, that associated with photosynthesis, along with down regulation of non-photochemical quenching (NPQ) genes and those involved in response to light stress. The developed ISR approach is applicable toward improving cultivation of diatomaceous production strains under high light intensity cultivation in indoor reactors, or in open outdoor ponds.

Fusion of Ferredoxin to Cytochrome P450 Enables Light-Driven Biosynthesis

- Conference: International Conference on Plant Synthetic Biology and Bioengineering
- Year: 2016
- Proceeding: International Conference on Plant Synthetic Biology and Bioengineering
- Group: General Submissions
- Session:

  Plastid Engineering: artificial photosynthesis, metabolites and nitrogen fixation

- Time:

  Saturday, December 17, 2016 - 12:00am-12:25am

Authors:
Jensen, P. E., University of Copenhagen

Plants produce a multitude of specialized metabolites with use in the medicinal, fragrance and flavor industries. Cytochromes P450 (P450s) are key enzymes in biosynthetic pathways of these metabolites[1]. P450s are difficult to express heterologously in active form, and due to their requirement for reducing power in the form of NADPH their use for in vitro and whole-cell production is complicated. We recently showed that plant P450s can be expressed in chloroplasts of tobacco plants and in cyanobacteria, where they will insert into the thylakoid membrane and that photosynthesis can support P450 catalytic activity independent of NADPH through the action ferredoxin [3,4]. In the current study, we report the fusion of ferredoxin with the plant P450 CYP79A1, which catalyzes the initial step of the pathway leading to biosynthesis of the cyanogenic glucoside dhurrin. Fusion with ferredoxin allows the cytochrome P450 enzyme to obtain electrons for catalysis directly from photosynthesis by interacting with photosystem I. Furthermore, the electrons captured by the fused ferredoxin domain are directed more effectively towards the P450 in competition with other ferredoxin requiring enzymes [2].

As a result, it partially overcomes the problem of competition for reduced ferredoxin by electron sinks coupled to endogenous metabolic pathways. The ferredoxin-P450 fusion enzyme obtains reducing power solely from its fused ferredoxin, but maintains a similar level of catalytic activity compared to unfused CYP79A1 at in vivo concentrations of soluble ferredoxin. This demonstrates that electron transfer from photosystem I to CYP79A1 is greatly accelerated as a consequence of the fusion. The fusion strategy reported here therefore forms the basis for increased partitioning of photosynthetic reducing power towards P450-dependent biosynthesis of important natural products. This approach has high potential for stably engineering cyanobacteria that enables high-level cytochrome P450-dependent production of high-value natural products in a light-driven manner.

References:

1. Lassen, LM, Zygadlo Nielsen A, Friis Ziersen BE; Gnanasekaran T, MÃ¸ller BL, **Jensen PE** (2014). Redirecting photosynthetic electron flow into light-driven synthesis of alternative products including high-value bioactive natural compounds. ACS Synth. Biol. 3(1): 1â€"12. DOI: 10.1111/ppl.12108

2. Mellor S, Nielsen A, Burow M, Motawia M, Jakubauskas D, MÃ¸ller BL, **Jensen PE** (2016). Fusion of ferredoxin and cytochrome P450 enables direct light-driven biosynthesis. ACS Chemical Biology, DOI: 10.1021/acschembio.6b00190.

3. Wlodarczyka A, Gnanasekarana T, Nielsen AZ, Zulu NN, Mellor SB, ThÃ¸fner JFB, Olsen CE, Mottawie MS, Burow M, Luckner M, Pribil M, Feussner I, MÃ¸ller BL, and **Jensen PE** (2016). Metabolic engineering of light-driven cytochrome P450 dependent pathways into Synechocystis sp. PCC 6803. Metabolic Engineering 33: 1-11. doi: 10.1016/j.ymben.2015.10.009.

4. Zygadlo Nielsen A**,** Friis Ziersen BE, Jensen K, Lassen LM, Olsen CE, MÃ¸ller BL, **Jensen PE** (2013). Redirecting photosynthetic reducing power towards bioactive natural product synthesis. ACS Synthetic Biology 2: 308â^'315. *dx.doi.org/10.1021/sb300128r.*

Genes Stacking of Multiple Traits to Improve Biomass for Biofuels

- Conference: International Conference on Plant Synthetic Biology and Bioengineering
- Year: 2016
- Proceeding: International Conference on Plant Synthetic Biology and Bioengineering
- Group: General Submissions
- Session:

  Engineering of Bioenergy traits

- Time:

  Sunday, December 18, 2016 - 12:00am-12:25am

Authors:
Aznar, A., Joint BioEnergy Institute, Lawrence Berkeley National Laboratory
Eudes, A., Joint BioEnergy Institute, Lawrence Berkeley National Laboratory
Vuu, K., Joint BioEnergy Institute, Lawrence Berkeley National Laboratory
Shih, P., Joint BioEnergy Institute, Lawrence Berkeley National Laboratory
Loqué, D., Joint BioEnergy Institute, Lawrence Berkeley National Laboratory
Chalvin, C., Joint BioEnergy Institute, Lawrence Berkeley National Laboratory
Lee, M. Y., Joint BioEnergy Institute
Gondolf, V. M., University of Copenhagen
Ebert, B., University of Melbourne
Birsdeye, D., Joint BioEnergy Institute, Lawrence Berkeley National Laboratory
Laursen, T., Joint BioEnergy Institute, Lawrence Berkeley National Laboratory
Scheller, H. V., Joint BioEnergy Institute, Lawrence Berkeley National Laboratory
Second-generation biofuels produced from biomass can help to decrease dependency on fossil fuels, which would have many economical and environmental benefits. To make biomass more suitable for biorefinery use we need a better understanding of plant cell wall biosynthesis. Increasing the ratio of C6 to C5 sugars in the wall is an important target for engineering of plants that are more suitable for downstream processing for

second-generation biofuel production. Likewise, decreasing the content of lignin is an important goal. We have studied the basic mechanisms of cell wall biosynthesis and identified genes involved in biosynthesis of pectic galactan including the GALS1 galactan synthase[1] and the URGT1 UDP-galactose transporter[2]. We have applied these findings to engineer plants that have a more suitable biomass composition and have developed synthetic biology and gene stacking tools[3] to achieve this goal. Plants were engineered to have up to three-fold increased content of pectic galactan in stems by expressing GALS1, URGT1 and a UDP-glucose epimerase. Furthermore, the increased galactan was engineered into plants that were already engineered to have low xylan content by restricting xylan biosynthesis to vessels where this polysaccharide is essential[4]. Finally, the high galactan and low xylan traits were stacked with low lignin obtained by expressing the *QsuB*gene encoding dehydroshikimate dehydratase[5]. By targeting the transgene expression to specific cell types, we could substantially improve saccharification while avoiding adverse effects on plant growth and development.

References

1. Liwanag, A.J., et al., Pectin biosynthesis: GALS1 in Arabidopsis thaliana is a beta-1,4-galactan beta-1,4-galactosyltransferase. Plant Cell, 2012. 24(12): p. 5024-36.

2. Rautengarten, C., et al., The Golgi localized bifunctional UDP-rhamnose/UDP-galactose transporter family of Arabidopsis. Proc Natl Acad Sci U S A, 2014. 111(31): p. 11563-8.

3. Shih, P.M., et al., A robust gene stacking method utilizing yeast assembly for plant synthetic biology. Nature Communications

4. Petersen, P.D., et al., Engineering of plants with improved properties as biofuels feedstocks by vessel-specific complementation of xylan biosynthesis mutants. Biotechnology for Biofuels, 2012. 5(1): p. 84.

5. Eudes, A., et al., Expression of a bacterial 3-dehydroshikimate dehydratase reduces lignin content and improves biomass saccharification efficiency. Plant Biotechnol J, 2015.

Engineering Functionally Specialized Monolignol 4-O-Methyltransferases to Alter Lignin Structure in Energy Crop

- Conference: International Conference on Plant Synthetic Biology and Bioengineering
- Year: 2016
- Proceeding: International Conference on Plant Synthetic Biology and Bioengineering
- Group: General Submissions
- Session:

  Engineering of Bioenergy traits

- Time:

  Sunday, December 18, 2016 - 12:25am-12:50am

Authors:
Liu, C. J.
Lignin is an irregularly crossâ€□linked biopolymer and primarily composed of *p*-hydroxyphenyl, guaiacyl, and syringyl subunits. Lignin composition dictates the degree of lignin condensation, reactivity, and thus the degradability of plant cell walls. Over the years we have been pursuing a strategy of etherifying the *para*-hydroxyl of the monolignols to disturb lignin biosynthesis.Â We reason that preferentially modifying the *para*-hydroxyl of particular lignin precursor to deprive its dehydrogenation propensity would prevent the formation of particular lignin subunits therefore may change lignin structure. Combining crystal structure determination with combinatorial active site saturation mutagenesis, we engineered a set of monolignol 4-*O*-methyltransferases that preferentially methylate the 4-hydroxyl of coniferyl alcohol, 5-hydroxylconiferyl alcohol, sinapyl alcohol, or ferulic acid, respectively. Crystal structure determination revealed that the engineered enzyme variants showed substantial alteration in their substrate-binding pocket that governs their substrate preferences. Expressing one of the engineered variants with preference for the phenolic moiety of syringyl lignin precursor in hybrid aspens, we

found that the action of this variant resulted in a significant alteration of lignin composition. The syringyl-lignin subunits and the "wall-bound" phenolics were reduced by more than 50% and, concomitantly, the cellulose fiber content was increased by up to 12%, compared to the control plants. These changes rendered a 62% increase in the simple sugar release upon saccharification of transgenic wood and up to 49% increase in ethanol yield, when the pre-treated and untreated woody biomass was subjected to the simultaneous saccharification and fermentation, compared to corresponding controls. Moreover, the greenhouse-grown transgenic plants showed no obvious biomass yield penalty. These data suggest that our strategy of etherifying lignin precursors by MOMT represents a useful biotechnological solution for effectively tailoring the digestibility of *Populus* woody biomass; the created aspens have promising biomass potential as feedstocks for biofuel production.

Metabolic Engineering Converts Sugarcane into an Oil Producing Crop

- Conference: International Conference on Plant Synthetic Biology and Bioengineering
- Year: 2016
- Proceeding: International Conference on Plant Synthetic Biology and Bioengineering
- Group: General Submissions
- Session:

  Engineering of Bioenergy traits

- Time:

  Sunday, December 18, 2016 - 12:50am-1:15am

Authors:
Altpeter, F., University of Florida - IFAS
Parajuli, S.
Karan, R., University of Florida - IFAS
Kannan, B., University of Florida - IFAS
Sanahuja, G.
Liu, H.
Shanklin, J.

Metabolic engineering to divert carbon flux from sucrose to oil in a high biomass crop like sugarcane has been proposed as a strategy to boost lipid yields per acre for biodiesel production. The energy content of plant oils in the form of triacylglycerols (TAGs) is two-fold greater compared to carbohydrates. However, vegetative plant tissues do not accumulate oil to a significant amount since fatty acid synthesis in these tissues serves primarily membrane construction, in addition TAGs undergo rapid turnover. Therefore, our objectives include metabolic engineering to:

1.) increase fatty acid synthesis,

2.) increase TAG synthesis from diacyl-glycerol and acyl-CoA

3.) optimize TAG storage

4.) minimize TAG hydrolysis in vegetative tissues.

Constitutive single or multiple gene expression/suppression cassettes were generated and co-delivered with the selectable *npt*II expression cassette by biolistic gene transfer into sugarcane callus. Plants were regenerated on geneticin containing culture medium and analyzed for presence and expression of target constructs by PCR and RT-PCR, respectively. Quantitative real PCR was performed to study the level of gene expression in transgenic TAG plants in different vegetative tissues. Plants were analyzed for TAG content by gas-chromatography and mass spectrometry (GC-MS).

TAG accumulation increased several hundred fold in vegetative tissues of transgenic sugarcane with co-expression/co-supression of multiple target genes. While events with low TAG accumulation co-expressed either fewer transgenes, or expressed the transgenes at lower levels. This research outcome will add value to the abundant sugarcane post-harvest residues for production of advanced biofuels.

Sugar Transport By Sweets: From Structures to Biosensors

- Conference: International Conference on Plant Synthetic Biology and Bioengineering
- Year: 2016
- Proceeding: International Conference on Plant Synthetic Biology and Bioengineering
- Group: General Submissions
- Session:

  Engineering of Bioenergy traits

- Time:

  Sunday, December 18, 2016 - 1:15am-1:40am

Authors:
Cheung, L., Carnegie Institution for Science
Frommer, W.
Chavez, T.
The SWEETs are a family of sugar transporters conserved in prokaryotes, plants and animals. In plants, they play key roles in numerous physiological processes including phloem loading, pollen nutrition, nectar secretion and pathogen susceptibility. We have performed extensive functional characterization of this family based on the atomic structures of their bacterial homologues (the SemiSWEET) and the vacuolar glucose transporter OsSWEET2b from rice. We take advantage of these atomic structures and a high throughput cell sorting approach to create a set of SWEET biosensors composed of transporter-fluorescent protein fusions. We use our biosensors to quantify the kinetics of the SWEETs, and to gain further insight into the molecular mechanisms underlying selectivity and gating.

Increasing the C6/C5 Sugar Ratio in Bioenergy Crops By Modulating Nucleotide Sugar Transporters and Glycosyltransferases

- Conference: International Conference on Plant Synthetic Biology and Bioengineering
- Year: 2016
- Proceeding: International Conference on Plant Synthetic Biology and Bioengineering
- Group: General Submissions
- Session:

  Engineering of Bioenergy traits

- Time:

  Sunday, December 18, 2016 - 1:40am-2:05am

Authors:
Scheller, H. V., Joint BioEnergy Institute, Lawrence Berkeley National Laboratory
Aznar, A., Joint BioEnergy Institute, Lawrence Berkeley National Laboratory
Ebert, B., University of Melbourne
Birdseye, D., Joint BioEnergy Institute, Lawrence Berkeley National Laboratory
Loqué, D., Joint BioEnergy Institute, Lawrence Berkeley National Laboratory
Chalvin, C., Joint BioEnergy Institute, Lawrence Berkeley National Laboratory
Laursen, T., Joint BioEnergy Institute, Lawrence Berkeley National Laboratory
Brandon, A. G., Joint BioEnergy Institute, Lawrence Berkeley National Laboratory,
Rautengarten, C., Joint Bioenergy Institute
Heazlewood, J.
Yan, J., Joint BioEnergy Institute, Lawrence Berkeley National Laboratory,
Biomass consists of about 30% of xylan, a polysaccharide composed of pentoses. Hexoses are more easily converted to biofuels and bioproducts, and therefore it is

advantageous to develop plants with a higher ratio of C6 to C5 sugars in their cell walls. We have identified more than 20 nucleotide sugar transporters responsible for transporting substrates into the Golgi lumen, where they are used by glycosyltransferases for synthesis of polysaccharides. The transporters we have identified transport ten different nucleotide sugars, including UDP-Gal, UDP-Rha, UDP-Xyl, UDP-GlcNAc, UDP-Araf, GDP-Man, and GDP-Fuc. By modulating the expression of UDP-Gal and UDP-Xyl transporters we have achieved an improved C6/C5 ratio in plant cell walls. Likewise, we have achieved a higher C6/C5 ratio by overexpressing a number of glycosyltransferases for polysaccharides rich in C6 sugars, or inhibiting enzymes responsible for synthesis of polysaccharides rich in C5 sugars. The resulting plants are indistinguishable from the wild type under normal growth conditions. Changing the cell walls of plants may lead to altered environmental resilience, and we have therefore tested the drought tolerance of some of the engineered plants. Surprisingly, many of the plants show increased drought tolerance.

A Toolkit of Programmable Salicylic Acid Regulated Transcription Factors to Engineer Plant Immune Responses

- Conference: International Conference on Plant Synthetic Biology and Bioengineering
- Year: 2016
- Proceeding: International Conference on Plant Synthetic Biology and Bioengineering
- Group: Poster Submissions
- Session:

  Poster Session

- Time:

  Friday, December 16, 2016 - 5:20pm-7:20pm

Authors:
de Lange, O., The University of Washington
Khakhar, A., Johns Hopkins University
Nemhauser, J., University of Washington
Klavins, E., University of Washington

Pathogen damage limits crop yields around the world but it can be controlled with interventions at the genetic level. Classical breeding has been successful at broadening the range of pathogen receptors encoded in crop genomes. We would like to use synthetic biology to rationally redesign the stages of the plant immune response that act downstream of these receptors.

This poster charts our development of synthetic transcription factors responsive to salicylic acid (SA), one of the key plant immune system hormones. SA is produced in response to pathogen receptor activation and is responsible for triggering transcriptional reprogramming. SA acts through a transcriptional co-activator NPR1. SA regulation of NPR1 occurs via multiple parallel mechanisms, and the downstream binding partners of NPR1 receive inputs from other hormone pathways, introducing cross-talk. Both the complex relationship between SA and NPR1 activity and the complex downstream crosstalk are hurdles to rational engineering.

We are currently porting the machinery for SA dependent protein degradation from plants into budding yeast so that we can use yeast as a simple chassis to characterise degradation dynamics. A minimal SA degron will then be fused to a nuclease-null Cas9 to direct transcriptional responses *in planta*.

Bicistronic Expression of Gibberellin 20-Oxidase and Secondary Wall Biosynthetic MYB Transcription Factor Improves Both Quality and Quantity of Woody Biomass Production in Poplar

ICPSBB
International Conference on Plant Synthetic Biology and Bioengineering

- Conference: International Conference on Plant Synthetic Biology and Bioengineering
- Year: 2016
- Proceeding: International Conference on Plant Synthetic Biology and Bioengineering
- Group: Poster Submissions
- Session:

  Poster Session

- Time:

  Friday, December 16, 2016 - 5:20pm-7:20pm

Authors:
Ko, J. H.
Cho, J. S.
Park, E. J., Korea Forest Research Institute
Choi, Y. I., Korea Forest Research Institute

Research of woody biomass production has increased as a potential renewable resource for biofuel. In this study, we present a novel approach for biotechnological improvement of the quantity and quality of woody biomass with minimizing undesirable growth penalties. We applied a multi-cistronic gene expression system, known as viral 2A peptide, which allows expressing two or more genes under the same promoter, thus capable to express two or more genes at the exactly same spatio-temporal manner. Our transgenic poplars designed to express both GA20-oxidease (a key enzyme to produce bioactive gibberellin (GA)) and a MYB transcription factor involved in secondary wall biosynthesis in developing xylem (DX) tissue-specific, resulted in a biomass increase up to 175%. Similar phenotypes were also observed in the transgenic Arabidopsis plants harboring the same gene construct. This phenotypic consequence was further verified in LMO-field experiment. During three months of actively growing season, our transgenic poplars grow up to 156% and 123% in the height and diameter, respectively, compared to wild-type poplar without any undesirable phenotypes. Our results demonstrate that the controlled production of GAs and secondary wall regulating MYB TF through a DX promoter can be utilized as an efficient biotechnological tool for producing enhanced plant biomass without undesirable side effects. This work was funded by the Basic Science Research Program through the National Research Foundation of Korea (NRF) (2011-0008840) and a grant from the Korea Forest Service (S111213L080110).

Bioproduction of Muconic Acid in Plants

- Conference: International Conference on Plant Synthetic Biology and Bioengineering
- Year: 2016
- Proceeding: International Conference on Plant Synthetic Biology and Bioengineering
- Group: Poster Submissions
- Session:

  Poster Session

- Time:

  Friday, December 16, 2016 - 5:20pm-7:20pm

Authors:
Vuu, K., Joint BioEnergy Institute, Lawrence Berkeley National Laboratory
Shih, P., Joint BioEnergy Institute, Lawrence Berkeley National Laboratory
Eudes, A., Joint BioEnergy Institute, Lawrence Berkeley National Laboratory
Loque, D., Joint BioEnergy Institute

The vast majority of bioengineering efforts for the production of commodity chemicals have been pursued in microorganisms, due to the ease of genetically manipulating and optimizing metabolic pathways in microbes such as *E. coli* and yeast. With improvements in plant synthetic biology tools and increasing interest in leveraging plants as a bioplatform for production of biologics or industrially-relevant chemicals, we explore the potential of using plant-based production of the chemical muconic acid â€“ an intermediate molecule that can be derived into several bioplastics. Most biological approaches have focused on producing muconic acid from a glucose feedstock in microbes. Here, we test and optimize various metabolic routes for the production of muconic acid within plants, enabling the direct production of muconic acid via photosynthesis. Plant-based metabolic engineering efforts may enable a more sustainable means of producing chemicals of interest and decrease our dependence on current practices that are heavily dependent on petroleum feedstocks.

Bistable Genetic Switch Design in Plant System

- Conference: International Conference on Plant Synthetic Biology and Bioengineering
- Year: 2016
- Proceeding: International Conference on Plant Synthetic Biology and Bioengineering
- Group: Poster Submissions
- Session:

  Poster Session

- Time:

  Friday, December 16, 2016 - 5:20pm-7:20pm

Authors:
Kassaw, T., Colorado State University
Zalewski, C., Front Range Biosciences
Schaumberg, K., Massachusetts Institute of Technology (MIT)
Antunes, M., Colorado State University
Prasad, A., Colorado State University
Medford, J. I., Colorado State University
Programming the behavior of biological traits using rationally designed genetic devices has been the goal of synthetic biology. Genetic switches are one of the devices essential for decision making at different phases of regulatory networks and allow programmable control over biological systems. To build robust biological based bistable switches, orthogonal genetic components and quantitatively determined input-output functions are vital. Here, we are reporting the characterization of genetic parts and the demonstration of a synthetic genetic toggle switch design in complex plant system. The circuit uses synthetic repressor-repressible promoter pairs that are quantitatively balanced and can be controlled by external inducers to toggle between two stable output states (*ON* vs. *OFF*). We have designed and quantitatively characterized the behavior of a series of these pairs transiently in protoplasts. The kinetic parameters derived from these assays were used to

guide the design of more complex genetic circuits for whole plant. So far, homozygous transgenic lines containing bistable genetic circuits are generated and showed reasonable toggle switch behavior under the control of external inputs. By modifying the output of these toggle switches, we hope to increase the engineering potential of useful plant traits.

Characterization of Arabidopsis Mutants with Minimal Level of Centromeric Histone: Impact on Chromosome Segregation and Induction of Haploid Plants

- Conference: International Conference on Plant Synthetic Biology and Bioengineering
- Year: 2016
- Proceeding: International Conference on Plant Synthetic Biology and Bioengineering
- Group: Poster Submissions
- Session:

  Poster Session

- Time:

  Friday, December 16, 2016 - 5:20pm-7:20pm

Authors:
Capitao, C., Gregor Mendel Institute
Akimcheva, S., Gregor Mendel Institute
Tanasa, S., CEITEC
Bulankova, P., Department of Plant Systems Biology, VIB
Raxwal, V., CEITEC
Mittelsten Scheid, O., Gregor Mendel Institute

Riha, K., CEITEC

Understanding fundamental principles of chromosome segregation machinery offers new opportunities for plant genome manipulation and engineering. The centromere-specific histone 3, CENH3, is an essential component of the centromere that determines assembly of functional kinetochores and hence, segregation and genetic transmissivity of chromosomes. Recently, a new method for induction of haploid plants based on modification of the centromeric histone has been described. This mechanism, called centromere-mediated genome elimination, is characterized by complete loss of a parental genome marked with a modified centromeric histone. This new method for plant haploidization can be used for rapid construction of recombinant inbred lines or generating complex mutants.

In this study we have isolated a novel allele (cenh3-2) of the Arabidopsis thaliana CENH3 gene that causes 100 fold reduction in the protein level. The cenh3-2 allele was found in a forward genetic screen aimed to identify mutants that can rescue infertility of plants deficient in proper completion of meiosis. This mutation leads to alternative splicing resulting in very low levels of functional transcript. Surprisingly, despite the drastic reduction of CENH3 in the centromeres, affected chromosomes are mitotically stable and exhibit only minor defects in meiotic segregation. This indicates that the functional centromere in plants may span a much smaller region than it has been postulated, facilitating the design of more efficient artificial chromosomes. Interestingly, cenh3-2 plants are very inefficient in haploid induction, indicating that the mechanism of genome elimination is not associated with low levels of CENH3 (weak centromere) but instead with alteration of CENH3 protein structure.

Our results are important for addressing the concept of the minimal centromere in plants and for understanding the mechanism of haploid induction.

Comparative Transcriptome Analysis of Qualea Grandiflora mart. Plants to Investigate the Role of Aluminum in This Cerrado Species Metabolism

- Conference: International Conference on Plant Synthetic Biology and Bioengineering
- Year: 2016
- Proceeding: International Conference on Plant Synthetic Biology and Bioengineering
- Group: Poster Submissions
- Session:

  Poster Session

- Time:

  Friday, December 16, 2016 - 5:20pm-7:20pm

Authors:
Cristina Costa e Silva, R., University of Brasilia

Title: Comparative transcriptome analysis of *Qualea grandiflora* Mart. plants to investigate the role of aluminum in this Cerrado species metabolism

Renata Cristina Costa e Silva*, Fernando Araripe Gonçalves Torres*, Gabriel da Rocha Fernandes, Simoni Campos Dias, Luiz Alfredo Rodrigues Pereira*.

*University of Brazilian, Postal Code 70910-90, Brazil.

Abstract. About 40% of agricultural land of the world is acid with high levels of aluminum (Al). Therefore, most tropical soils are acid. Soil acidity occurs through natural processes that result in the accumulation Al species, especially Al^{3+}. Cerrado plants species of Vochysiaceae family, such as *Qualea grandiflora*, accumulate Al and require it for proper growth and development. Based on that, the transcriptome profiles of leaves from 8-month old plants grown with and without Al were investigated. *De novo* assembly of individual transcriptome followed by sequence clustering yielded 134,494 non redundant sequences, from which 16,719 transcripts were differentially expressed between the two treatments. Kyoto Encyclopedia of Genes and Genomes (KEGG) enrichment analyses indicated that 52 and 58 orthology groups, were overrepresented in upregulated transcripts in control and Al-treated plants, respectively. A total of 400 transcription and transduction factors and 208 transporter genes were differentially regulated in *Q. grandiflora* leaves in response to Al. Some genes involved in carbohydrate metabolism, environmental adaptation, resistance to abiotic and biotic stresses, as well as pathogen defense were highly expressed in leaves of Al-treated plants, implicating them in Al metabolism in this plant species. Moreover, in the absence of Al, the most regulated genes were related to stress response mechanism such as programed cell death, production of secondary metabolites (terpenes) and antioxidant activity. The transcriptome also revealed a gene network in the leaves regulated by Al and provided valuable information on Al associated genes. Additionally, a detailed insight into the molecular mechanisms associated with the role of Al in *Q. grandiflora* may provide future applications to crop improvement.

key words: transcriptome, Al-accumulating, metabolism.

Biography

Renata Cristina Costa e Silva is a PhD student at the Department of Botany, University of Brazilian. Her thesis proposal is directed towards unravelling the metabolic role of Al in Cerrado plants. These plants need Al to grow and develop. This research has been funded by FAPDF (Research funding foundation of Brasilia) and CAPES (Coordination for the Improvement of Higher Education Personnel).

Presenting author details

Full name: Renata Cristina Costa e Silva
Contact number: 55(62) 9122-3167
Email adress: renatacrisbio@yahoo.com.br
Category: Poster presentation

Comprehensive Analysis of Heavy Metal-Associated Domain Containing Genes Using Nitric Oxide-Mediated Transcriptomic Analysis in Arabidopsis

ICPSBB International Conference on Plant Synthetic Biology and Bioengineering

- Conference: International Conference on Plant Synthetic Biology and Bioengineering
- Year: 2016
- Proceeding: International Conference on Plant Synthetic Biology and Bioengineering
- Group: Poster Submissions
- Session:

  Poster Session

- Time:

  Friday, December 16, 2016 - 5:20pm-7:20pm

Authors:
Imran, Q. M., Kyungpook National University
Falak, N.
Hussain, A.
Mun, B. G.
Yun, B. W.
Metal ions play important roles in cellular and subcellular processes in both animals and plants. Heavy metal associated domain (HMA) containing genes encode small identical proteins of 69 to 73 amino acid in length mostly associated with transport of metal ions in and out of the cell. They are very important for spatio-temporal transportation of metal ions for binding with various enzymes and co-factors. We identified 16 different Arabidopsis HMA genes differentially expressed in response to nitrosative stress induced

by infiltrating 1 mM CysNO (S-nitrocysteine) into the leaf apoplast. Out of the 14 genes, expression of 8 HMA genes was increased, whereas, the expression of 6 genes was reduced. Computational characterization of HMA genes from Arabidopsis revealed the presence of a highly conserved *Cys-XX-Cys* motif within the HMA domain which may be target for S-nitrosylation and trans-nitrosylation. Promoter analysis also revealed the presence of important *Cis*-regulatory elements within a region 1.5 kb upstream of the transcriptional initiation site such as TCA element, CAAT, TATA-box. Further characterization of HMA gene through functional genomics showed that *at1g51090* plants were resistant to the bacterial pathogen *Pseudomonas syringae*(DC3000). Additionally, this gene positively regulates shoot development while negatively regulates root development under oxidative stress (induced by H_2O_2 and Methyl Vilogen) and nitrosative stress conditions induced by (CysNO and GSNO). Overall, these results imply that HMA domain containing genes may play important role in plant development and immunity.

•

Crop-Biorefinery System for Low-Cost Advanced Renewable Fuel and Chemicals

- Conference: International Conference on Plant Synthetic Biology and Bioengineering
- Year: 2016
- Proceeding: International Conference on Plant Synthetic Biology and Bioengineering
- Group: Poster Submissions
- Session:

  Poster Session

- Time:

  Friday, December 16, 2016 - 5:20pm-7:20pm

Authors:
Oikawa, A.

Sustainable ecosystems and U.S. energy security depend on more efficient fuel and chemical production from renewable sources. Demand for renewable hydrocarbon fuels and chemicals is rapidly growing and is expected to continue rising for several decades. However, current biorefineries face cost competitiveness challenges due to inadequate yield and quality of biomass and lack of robust fermentation hosts and product diversity to maximize income compared to petrochemical counterparts. Fortunately, improved energy feedstock crops for biomass and non-conventional oleaginous yeasts for conversion to desired products have recently emerged as candidates for metabolic engineering and/or synthetic biology approach that can significantly enhance the economics for production of biodiesel, jet fuels, and advanced high-value chemicals. We at AFINGEN and Joint Bioenergy Institute are developing synthetic biology methods to enable host eukaryotic organisms (plants, fungi, and yeast) to improve yield and composition of commercially attractive products. Among a suite of synthetic biology methods, AFINGEN has licensed a patented technology, the Artificial Positive Feedback Loop or â€œAPFLâ€, from the University of California. The APFL technology offers a robust path to produce target biofuels from cellulosic derived sugars with minimal cis-genetic manipulation and improved genetic stability compared to conventionalÂ bio-engineering. The outcome demonstrates a major breakthrough in synthetic biology that could lead to significantly higher yields and more diverse hosts that can be utilized in the proven, industrial-scale processes of fermentation. By multiplying the amplification and/or reduction of target compounds with unprecedented specificity and improved tolerance, engineered fuel crops may allow high-yield of biomass, more cellulose, and less lignin with healthy robust growth. The increased quantity of higher quality feedstocks provide the nutrient substrate for more productive engineered yeasts that produce much higher volumes and much lower production costs of currently scarce or difficult to extract compounds.

Design, Synthesis, and Testing Toward a 57-Codon Genome

- Conference: International Conference on Plant Synthetic Biology and Bioengineering
- Year: 2016
- Proceeding: International Conference on Plant Synthetic Biology and Bioengineering
- Group: Poster Submissions
- Session:

  Poster Session

- Time:

  Friday, December 16, 2016 - 5:20pm-7:20pm

Authors:
Ostrov, N., Harvard Medical School
Recodingâ€"the repurposing of genetic codonsâ€"is a powerful strategy for enhancing

genomes with functions not commonly found in nature. Here, we report computational

design, synthesis, and progress toward assembly of a 3.97-megabase, 57-codon Escherichia

coli genome in which all 62,214 instances of seven codons were replaced with synonymous

alternatives across all protein-coding genes. We have validated 63% of recoded genes

by individually testing 55 segments of 50 kilobases each.We observed that 91% of tested essential genes retained functionality with limited fitness effect. We demonstrate identification and correction of lethal design exceptions, only 13 of which were found in 2229 genes. This work underscores the feasibility of rewriting genomes and establishes a framework for large-scale design, assembly, troubleshooting, and phenotypic analysis of synthetic organisms.

Dominant Suppression of Xylan Biosynthesis Using IRX10

- Conference: International Conference on Plant Synthetic Biology and Bioengineering
- Year: 2016
- Proceeding: International Conference on Plant Synthetic Biology and Bioengineering
- Group: Poster Submissions
- Session:

  Poster Session

- Time:

  Friday, December 16, 2016 - 5:20pm-7:20pm

Authors:
Brandon, A. G., Joint BioEnergy Institute, Lawrence Berkeley National Laboratory,
Dominant suppression of xylan biosynthesis using IRX10

Andrew Brandon[1,2], Mi Yeon Lee[1], Henrik Vibe Scheller[1,2]

[1]Joint Bioenergy Institute and Biological Systems and Engineering Division, Lawrence Berkeley National Laboratory, Berkeley, CA 94720, USA

[2]Department of Plant and Microbial Biology, University of California, Berkeley, CA 94720, USA

The Î^{2}-(1,4)-linked xylose homopolymer xylan comprises 20-30% of the biomass of dicot plant species and nearly half the biomass of grasses. It is the most abundant polysaccharide after cellulose in most plants and, thus, one of the most abundant biopolymers on earth. While its role in the cell wall is not entirely known, it is believed to coat and/or crosslink cellulose microfibrils. Mutants in xylan biosynthetic genes have thinner and weaker cell walls. The vessels of these mutants tend to collapse under the negative pressure of water transport, causing a severely dwarfed whole-plant phenotype. Despite its abundance in the cell wall and importance to plant health, the composition of xylan makes it unfavorable for the conversion of plant biomass to biofuels and bioproducts. The xylan backbone is composed entirely of the 5-carbon sugar xylose. Microbial fermentation of 5-carbon sugars is relatively inefficient, and the presence of 5-carbon sugars inhibits the fermentation of abundant 6-carbon sugars like glucose. Additionally, the xylan backbone is heavily substituted with acetyl groups. The released acetate creates a toxic environment for microbial fermentation. So, any engineering efforts that can reduce the amount of xylan in plant biomass can have significant positive impacts on the viability and cost-efficiency of advanced biofuels. Transformation of plants with dominant alleles is an ideal method to engineer a range of bioenergy crops, especially since few, if any, xylan mutants of biotechnologically relevant crops exist. We have therefore explored the possibility of developing dominant supressors of xylan biosynthesis. While xylan biosynthesis is still poorly understood, one pair of homologous enzymes *Irregular xylem (IRX) 10* and *10-like* has unambiguously demonstrated xylan synthase activity. We have been able to dramatically reduce or abolish this catalytic activity by identifying and mutating highly conserved residues in the IRX10 amino acid sequence. Then, we were able to out-compete the native IRX10 enzyme in a wild-type background by constitutively overexpressing the mutant form, the effect being to suppress xylan biosynthesis and reduce the amount of xylan in the plant to varying degrees.

Engineering of Crassulacean Acid Metabolism (CAM) Modules into C3 Species Via Synthetic Biology Approach

- Conference: International Conference on Plant Synthetic Biology and Bioengineering
- Year: 2016
- Proceeding: International Conference on Plant Synthetic Biology and Bioengineering
- Group: Poster Submissions
- Session:

  Poster Session

- Time:

  Friday, December 16, 2016 - 5:20pm-7:20pm

Authors:
Hu, R., Oak Ridge National Laboratory
Liu, D., Oak Ridge National Laboratory
Borland, A., Newcastle University
Tuskan, G., Oak Ridge National Laboratory
Yang, X., Oak Ridge National Laboratory
Crassulacean acid metabolism (CAM) is a specialized photosynthetic pathway that enhances water use efficiency (WUE) in CAM plants in comparison with C_3 and C_4 plants. The CAM biochemistry features two temporally-separated modules: 1) a carboxylation module for nocturnal CO_2 fixation and accumulation of malic acid in the vacuole; and 2) a decarboxylation module for release of CO_2 from malate during the daytime. From the genomics resources available for two obligate CAM species *Agave americana* (monocot) and *Kalanchoë laxiflora* (eudicot), we identified the key genes involved in the CAM carboxylation and decarboxylation processes. Multiple genetic

circuits were created for the CAM carboxylation and decarboxylation modules using synthetic biology approach. These genetic circuits were introduced into two model plants species (i.e., *Arabidopsis* and tobacco) for testing and optimization of the circuit design. Furthermore, the CAM genetic circuits are being transformed into poplar, an important C_3 bioenergy crop. This research has a great potential for accelerating the genetic improvement of WUE and drought tolerance in bioenergy and food crops.

Genome-Wide Study on Cultivar-Specific and Stripe Rust Responsive Mirnas in Triticum aestivum

ICPSBB
International Conference on Plant Synthetic Biology and Bioengineering

- Conference: International Conference on Plant Synthetic Biology and Bioengineering
- Year: 2016
- Proceeding: International Conference on Plant Synthetic Biology and Bioengineering
- Group: Poster Submissions
- Session:

  Poster Session

- Time:

  Friday, December 16, 2016 - 5:20pm-7:20pm

Authors:
Ramachandran, S.

Wheat represents approximately 30% of the worldâ€™s grain production with more than 200 million hectares under cultivation. Productivity of the crop is mainly governed by genetic and environmental factors influencing growth and development. Understanding the genetic factors that regulate these processes will help in developing varieties with better yield potential and disease resistance. A major regulartor of eukaryotic gene expression are microRNAs that control myriad of physiological processes from development to plant stress responses. Using 12 small RNA libraries prepared from two wheat cultivars infected with stripe rust fungus (*Puccinia striiformis*), we identified 43 previously-known miRNAs, 50 variants of known miRNAs and 93 candidate novel miRNAs. RNA sequencing revealed 20 cultivar-specific miRNAs, 6 miRNAs differentially expressed between the two cultivars and 4 miRNAs responsive to stripe rust infection. These results were validated using RT-PCR and qRT-PCR. Using different target prediction algorithms,145 miRNAs were predicted to target wheat genes, while 69 miRNAs were found to target fungal genes. Of the fungal targets a majority coded for small, secreted, cellular proteins, suggesting a trans-kingdom regulation of gene expression potentially involved in the cultivar-specific resistance to stripe rust. Overall, this study contributes to the current repository of wheat miRNAs and provides novel information on the yet uncharacterized roles for miRNAs in wheat host-pathogen interactions.

In Situ engineering of Bacterial Communities

- Conference: International Conference on Plant Synthetic Biology and Bioengineering
- Year: 2016
- Proceeding: International Conference on Plant Synthetic Biology and Bioengineering
- Group: Poster Submissions
- Session:

  Poster Session

- Time:

  Friday, December 16, 2016 - 5:20pm-7:20pm

Authors:
Brophy, J. A. N.
Grossman, A. D.

Plants are colonized by complex microbial communities that can promote plant growth, increase disease resistance, and improve environmental stress tolerance. The bacteria associated with plants are often difficult to manipulate genetically, which limits scientific studies and microbiome engineering. We constructed a versatile *Bacillus subtilis* donor strain to deliver an engineered Integrative Conjugative Element (ICE*Bs1*) to complex communities of bacteria *in situ*. We demonstrate that ICE*Bs1* can be used to deliver heterologous DNA to >17 Gram-positive bacterial species isolated from human and soil microbiomes (*e.g., Bacillus* sp., *Paenibacillus* sp., *Enterococcus* sp., *Staphylococcus* sp.). Then, ICE*Bs1* is used to create synthetic diazatrophs by transforming soil bacteria with nitrogenase (nif) gene clusters.

Inter-Specific Differences in Growth, Leaf Phenology in Seedlings of Cork Oak (Quercus suber L.), Zeen Oak (Quercus canariensis Willd.) and Their Hybrid Afares Oak (Quercus afares Pomel) Grown in the Nursery

ICPSBB
International Conference on Plant Synthetic Biology and Bioengineering

- Conference: International Conference on Plant Synthetic Biology and Bioengineering
- Year: 2016
- Proceeding: International Conference on Plant Synthetic Biology and Bioengineering
- Group: Poster Submissions
- Session:

  Poster Session

- Time:

  Friday, December 16, 2016 - 5:20pm-7:20pm

Authors:
Mhamdi, S.

Outside of the systematic, Leaf Life Span (LLS) is used to classify trees into two main groups: evergreen and deciduous species. It varies according to the forms of life between taxonomic groups. Co-occurrence of deciduous and evergreen oaks is common in some Mediterranean type climate areas. Nevertheless, in the Tunisian forests, there is no enough information about the functional inter-specific diversity among oak species, especially in the mixed stand marked by the simultaneous presence of *Q. suber* L, *Q.canariensis Willd* and their putative hybrid (*Q. afares*), endemic oak species threatened with extinction. This study has been conducted to estimate the LLS, the relative growth rate and the count of different growth flushes in samplings in semi-controlled conditions. Our study took 17 months, with an observation's interval of 4 weeks. The aim is to characterize and compare the hybrid species to the parental ones. Differences were observed between species, both at growth and phenology. Indeed, *Q. suber* saplings reached higher total height and number of growth flushs then *Q.canariensis*. On the other hand, *Q. afares* showed much less growth flushs and is the less performing species. The LLS of parental species has exceeded the duration of the experiment, but their hybrid lost all leaves on all cohorts. The short LLS hybrid species are in accordance with his phenology in the field but for *Q. canariensis* the long LLS contrast with observations in the field where phenology is strictly annual. This study allowed us to differentiate the hybrid from both parental species at seeding stage.

•

Lectin Receptor Protein Kinase Is Induced in Plant Roots in Response to the Endophytic Fungus, Piriformospora indica

- Conference: International Conference on Plant Synthetic Biology and Bioengineering
- Year: 2016
- Proceeding: International Conference on Plant Synthetic Biology and Bioengineering
- Group: Poster Submissions
- Session:

  Poster Session

- Time:

  Friday, December 16, 2016 - 5:20pm-7:20pm

Authors:
Lal, N., Jamia Hamdard, New Delhi
Abdin, M. Z.
Verma, P.
Upadhyaya, K. C.
The symbiotic association of *Piriformospora indica, a member of the order Sebacinales,* provides growth promotion leading to increased biomass as well as enhanced plant tolerance to biotic and abiotic stresses in a variety of plant species. Since various protein kinases are thought be involved in mutual signaling in symbiotic interactions, we analyzed transcript profile of rice protein kinase genes in the infection process by real-time PCR. A rice L-type lectin protein kinase (*OslecRK*) and *Arabidopsis AtLecRK* genes are highly upregulated in roots during the initial stage of symbiotic association. In order

to investigate the role of *LecRK* gene in the root colonization, *Arabidopsis* mutant line (*Atlecrk*) of rice homolog was used which shows the restricted intracellular penetration of the *P. indica*. The results show that LecRK might be a crucial factor in perception and recognition during plant-*P. indica* association by regulating fungal invasion in the host root cells.

Lipid Droplets As a New Platform for Protein Expression in Chlamydomonas Reinhardtii

ICPSBB
International Conference on Plant Synthetic Biology and Bioengineering

- Conference: International Conference on Plant Synthetic Biology and Bioengineering
- Year: 2016
- Proceeding: International Conference on Plant Synthetic Biology and Bioengineering
- Group: Poster Submissions
- Session:

  Poster Session

- Time:

  Friday, December 16, 2016 - 5:20pm-7:20pm

Authors:
Bakowski, K., University of Copenhagen
Jensen, P. E., University of Copenhagen
Nielsen, A. Z., University of Copenhagen

Lipid droplets (LDs) are most often associated with biofuel production, as they are rich in neutral lipids, a desirable substrate for biodiesel production. The highly hydrophobic core of the LDs function as a storage compartment of variable small compounds, like carotenoids, sterols and terpenoids. They also have a physiological function in transporting lipids and proteins between membranes as well as storing them in the response to variable environmental stresses [1].

With the goal to sustainably produce high value products in the green algae *Chlamydomonas reinhardtii* we aim at production of heterologous proteins and assembling metabolic pathways on the surface of LDs by fusing enzymes to the anchoring proteins i.e. oleosins. Oleosins were used already for production of recombinant proteins in E. coli, yeasts and tobacco but not yet for expression of active enzymes and not yet in green eukaryotic algae. Oleosins are relatively small proteins with a highly hydrophobic proline-knot motif which anchor them to the LDs, and most probably, their insertion is driven solely by their physicochemical properties. By this strategy we will facilitate spatial organization of the enzymes and secretion of the hydrophobic products to the interior of the LDs [2].

First to show proper functioning of anchoring proteins we fused a fluorescent protein to different oleosins and successfully targeted these reporters to LDs in *C. reinhardtii.* The specific localization and bright fluorescence of these fusion proteins allows for detailed tracking of LDs and single-cell analysis of LDâ€™s physiology. By use of image cytometry we can assess the number, size and distribution of LDs in every cell, and compare the distributions between strains containing different oleosin-reporter constructs at different growth stages. Next we will try to express cytochrome P450s from the dhurrin pathway from *Sorhum bicolor* by exchanging the native transmembrane region with the LD anchoring proteins. Enzymes from the dhurrin pathway form functional biosynthetic protein complexes, so-called metabolons, facilitating shuttling of intermediates between enzymes and increasing formation of the final product. Our aim is to co-localize enzymes from this pathway and investigate if the metabolon formation still takes place. This proof-of-concept will be a good starting point for assembly of more complicated biosynthetic pathways [3, 4].

1. Thiam A. R., Farese R. V., Walther T. C. The biophysics and cell biology of lipid droplets. Nature Reviews Molecular Cell Biology 14: 775-786 (2013).
2. Bhatla S. C., Kaushik V., Yadav M. K. Use of oil bodies and oleosins in recombinant protein production and other biotechnological applications. Biotechnology Advances 28: 293â€“300 (2010).
3. Laursen T., MÃ¸ller B. L., Bassard J-E. Plasticity of specialized metabolism as mediated by dynamic metabolons. Trends in Plant Science 20: 20-32 (2015).
4. Nielsen-Zygadlo A., Mellor S. B., Vavitsas K., Wlodarczyk A. J., Gnanasekaran T., Ramos H. de Jesus M. P., King B. C., Bakowski K., Jensen P. E. Extending the biosynthetic repertoires of cyanobacteria and chloroplasts. The Plant Journal 87: 87â€“102 (2016).

Metabolic Profile of Qualea Grandiflora mart. (Vochysiaceae) Reveals the Importance of Aluminium in This Plant Metabolism

- Conference: International Conference on Plant Synthetic Biology and Bioengineering
- Year: 2016
- Proceeding: International Conference on Plant Synthetic Biology and Bioengineering
- Group: Poster Submissions
- Session:

  Poster Session

- Time:

  Friday, December 16, 2016 - 5:20pm-7:20pm

Authors:
Melo, J. R.

JÃ©ssica JÃ©ssica 2 7 2016-09-12T17:26:00Z 2016-09-12T17:40:00Z 1 435 2354 19 5 2784 14.00 Clean Clean false 21 false false false PT-BR X-NONE X-NONE /* Style Definitions */ table.MsoNormalTable {mso-style-name:"Tabela normal"; mso-tstyle-rowband-size:0; mso-tstyle-colband-size:0; mso-style-noshow:yes; mso-style-priority:99; mso-style-parent:""; mso-padding-alt:0cm 5.4pt 0cm 5.4pt; mso-para-margin-top:0cm; mso-para-margin-right:0cm; mso-para-margin-bottom:10.0pt; mso-para-margin-left:0cm; line-height:115%; mso-pagination:widow-orphan; font-size:11.0pt; font-family:"Calibri","sans-serif"; mso-ascii-font-family:Calibri; mso-ascii-theme-font:minor-latin; mso-hansi-font-family:Calibri; mso-hansi-theme-font:minor-latin; mso-fareast-language:EN-US;}

Title: Metabolic profile of *Qualea grandiflora* Mart. (Vochysiaceae) reveals the importance of aluminium in this plant metabolism.

Jessica Regina Melo*, Renata Cristina Costa e Silva, Thomas Christopher Rhys Williams, JoÃ£o Pedro Scheidemantel ConceiÃ§Ã£o Silva, Luiz Alfredo Rodrigues Pereira.

*University of Brazilian, Postal Code 70910-90, Brazil.

Abstract: The trivalent aluminum ion (Al^{+3}) is highly toxic and a major cause of stress to sensitive plants in acid soils. However, many species are well adapted to conditions of acid soils and high Al concentrations. Plant Al-adaptation mechanisms are classified as following: a) resistance, where Al is excluded by the root system preventing its entrance into the plant, and; b) tolerance where Al is taken up by the plant and its toxic effects are neutralized internally. Moreover, some native species from the Cerrado, besides accumulating Al, exhibit apparent deficiency symptoms in its absence. Therefore, the objective of this work was to study the physiological and metabolic changes in *Qualea grandiflora* Mart. grown with and without Al, as well as to analyze the effects of the absence of Al on chloroplast ultrastructure. Metabolite profiling of leaves together with plant growth analysis and measurement of photosynthetic pigments from plants of both treatments was therefore carried out. Furthermore, we conducted a comparison of chloroplast ultrastructure in *Q. grandiflora* leaves taken from plants grown with and without Al. Al-supplemented plants grew better than those of the control and had higher contents of chlorophylls (*a* e *b*) and carotenoids (24.73, 8.27 and 6.20 $mg.cm^{-2}$). The ultrastructural analysis revealed that the absence of Al during plant development led to progressive disintegration of chloroplasts. The metabolic profiles showed a relationship between levels of organic acids (citrate and oxalate) and Al in leaf tissue, suggesting the involvement of these acids in Al-chelation. The results of this study indicate that Al plays an important role in *Q. grandiflora* metabolism.

Keywords: Al-accumulating, Vochysiaceae, chloroplasts, organic acids

Biography

Jessica Regina Melo is a masterâ€™s degree student at the Department of Botany, University of Brazilian. Her thesis proposal is directed towards analyzing the effect of the Al absence in accumulating plants from Cerrado, and unravelling the tolerance mechanism of these plants. This research has been funded by FAPDF (Research funding foundation of Brasilia) and CAPES (Coordination for the Improvement of Higher Education Personnel).

Presenting author details

Full name: Jessica Regina Melo

Contact number: 55(61) 98211-9810

Email adress: jessicamelo2010@yahoo.com.br

Category: Poster presentation

Phosphatase from the Haloacid Dehalogenase (HAD) Superfamily Catalyze the Elusive Dephosphorylation Step of Riboflavin Biosynthesis in Arabidopsis thaliana Chloroplasts

- Conference: International Conference on Plant Synthetic Biology and Bioengineering
- Year: 2016
- Proceeding: International Conference on Plant Synthetic Biology and Bioengineering
- Group: Poster Submissions
- Session:

  Poster Session

- Time:

  Friday, December 16, 2016 - 5:20pm-7:20pm

Authors:
Sa, N.

Flavin nucleotides (FMN and FAD) are essential cofactors that are involved in numerous redox reactions and other biological processes including phototropism, photorepair of DNA and circadian clocks. Riboflavin (vitamin B2) is the direct precursor of flavin nucleotides and its biosynthesis in plants has not been fully elucidated. The long-term goal of our research is to advance fundamental understanding of plant flavin metabolism, and to enable crop enrichment in vitamin B2. In plants, enzymes catalyzing all but one step on the riboflavin biosynthesis pathway have been identified. The dephosphorylation step has been a long-standing gap on this pathway, because the enzymes hydrolyzing 5-amino-6-ribitylamino-2,4(1*H*,3*H*) pyrimidinedione 5â€™-phosphate (ARPP) are not well understood. The goal of this work is to identify and characterize the enzyme that catalyzes the dephosphorylation reaction on riboflavin biosynthesis pathway, from the model plantÂ *Arabidopsis thaliana.Â* Towards that goal, we cloned and purified eight candidate ARPP phosphatases belonging to Haloacid Dehalogenase (HAD) Superfamily. *In vitro* screening found two ARPP phosphatases (AtcpFHy/PyrP1 and AtPyrP2) catalyze the dephosphorylation reaction. Plastidial subcellular localization of AtcpFHy/PyrP1 and AtPyrP2 were confirmed and their biochemical features were studied. The physiological roles of AtcpFHy/PyrP1 and AtPyrP2 on flavin metabolism and plants growth were studied by modifying the target gene expression. T-DNA knockout of AtcpFHy/PyrP1 did not affect the flavin profile of the transgenic plants, whereas suppression of AtPyrP2 reduces flavin content and affects plant growth.

Plant Foundry: Biopolymers, Biopharmaceuticals and Medically Effective Secondary Metabolites (Vaccines, antibodies, and metabolites)

ICPSBB
International Conference on Plant Synthetic Biology and Bioengineering

- Conference: International Conference on Plant Synthetic Biology and Bioengineering
- Year: 2016
- Proceeding: International Conference on Plant Synthetic Biology and Bioengineering
- Group: Poster Submissions
- Session:

  Poster Session

- Time:

  Friday, December 16, 2016 - 5:20pm-7:20pm

Authors:
Scott, E., Atlantic International University

Abstract

Biopolymers in drug delivery, a polymeric substance occurring in livingorganisms. Polymers
Chemical approaches. Profile of biopolymersand biopharmaceuticals. Identification of medically
effective secondary metabolites; such as, vaccines, antibodies, and metabolites. All aspects of
plant foundry of plant synthetic biology will be discussed with poster presentation specifically

pertainingto and focusing on biopolymers and biopharmaceuticalsand their medicinal effect and
portal oftravel via Physician disbursement.

Key Words: Biopolymers, Biopharmaceuticals, Medica||yEffective, Secondary Metabolites,
Vaccines, Antibodies, Metabolites.

Plant Pathway Discovery and Bioengineering of the Most Popular Flavor, Vanillin

- Conference: International Conference on Plant Synthetic Biology and Bioengineering
- Year: 2016
- Proceeding: International Conference on Plant Synthetic Biology and Bioengineering
- Group: Poster Submissions
- Session:

  Poster Session

- Time:

  Friday, December 16, 2016 - 5:20pm-7:20pm

Authors:
Gallage, N.

Plant pathway discovery and bioengineering of the most popular flavor, Vanillin

Nethaji J Gallage and Birger Lindberg Møller

Plant Biochemistry Laboratory, Department of Plant and Environmental Sciences, University of Copenhagen

Vanilla and its key flavour component vanillin, is a universally appreciated flavour and probably the most popular plant natural product, being derived from the seedpods of the orchid *Vanilla planifolia* and other related Vanilla species. Vanillin (3-methoxy-4-hydroxybenzaldehyde) is the main flavor compound in the vanilla extract and is the basis and an additive of sweets, ice creams, soft drinks and many more products in the food, beverage and pharmaceutical industry. Our studies during last 4 years reveal how the vanilla orchid produces the most popular aroma compound in the world. In the vanilla pod, vanillin synthesize via ferulic acid and its glucoside. A single hydratase/lyase type enzyme designated vanillin synthase (*Vp*VAN) catalyzes direct conversion of ferulic acid and its glucoside into vanillin and its glucoside, respectively[1].

In 2010, the annual global sales of vanillin reached more than 15,000 tons. Nowadays, it is less than 1% of the global production of vanillin is derived from vanilla pods while the majority is produced synthetically using e.g. lignin and eugenol as starting materials. Production of 1 kilogram (kg) of vanillin requires approximately 500 kg of vanilla pods corresponding to the hand pollination of approximately 40,000 vanilla orchid flowers. This is why; a huge surge in the exploration of more environmentally friendly biosynthetic procedures to make natural flavours has risen[2]. Moreover, Industrial application of bioengineered microorganisms for vanillin production has gained quite a lot of attention not only from the flavour and fragrance industries, but also from environmental groups and the general public[3, 4]. The recent identification of *Vp*VAN from the vanilla orchid can contribute to an entirely new opportunity for biotechnology based production of natural vanillin[2]. We currently in the process of constructing an exclusive vanilla orchid vanillin production in engineered yeast encompassing vanilla orchid vanillin biosynthetic genes.

Furthermore we have initiated and involved in a project aims at unlocking the potential in Vanilla farming in Africa. Despite being the Worldâ€™s favorite flavor, Vanilla and vanilla production faces many challenges. Out of 110 Vanilla species, only three have been cultivated for farming, all of South-American origin, but grown in non-native environments primarily in Madagascar and regions of Africa. Apart from reducing the species gene-pool markedly and thus leaving the cultivated species and production reliability highly vulnerable to pests and diseases, the current production is highly labor demanding requiring intensive care and hand pollination as there are no natural pollinators in these non-native environments. However, in the African continent several native Vanilla species are found, but none have been elucidated for cultivability due to lack of resources. This project is the first comprehensive scientific study ever initiated on characterizing metabolite profile of 100 vanilla species with the final goal of discovering a vanilla species with African origin that has high quality vanilla pods and is better adapted to the African continent with possible natural pollinators in Uganda.

1. Gallage, N.J. et al. Vanillin formation from ferulic acid in Vanilla planifolia is catalysed by a single enzyme. *Nat Commun* **5** (2014).

2. Gallage, N.J. & MÃ¸ller, B.L. Vanillinâ€"Bioconversion and Bioengineering of the Most Popular Plant Flavor and Its De Novo Biosynthesis in the Vanilla Orchid. *Molecular Plant* **8**, 40-57 (2015).

3. Waltz, E. Engineers of scent. *Nature biotechnology* **33**, 329â€"332 (2015).

4. Bomgardner, M.M. The problem with vanilla, After vowing to go natural, food brands face a shortage of the favored flavor. *Chemical and engineering news* **94**, 38-42 (2016).

Precision Genome Editing in Sugarcane

- Conference: International Conference on Plant Synthetic Biology and Bioengineering
- Year: 2016
- Proceeding: International Conference on Plant Synthetic Biology and Bioengineering
- Group: Poster Submissions
- Session:

  Poster Session

- Time:

  Friday, December 16, 2016 - 5:20pm-7:20pm

Authors:
Oz, T., University of Florida - IFAS
Karan, R., University of Florida - IFAS
Altpeter, F., University of Florida - IFAS
Genome editing with a designer nuclease such as CRISPR/Cas9 enabling precise targeting and introduction of double stranded DNA breaks has been described in various plant genomes. Subsequent cellular repair mechanisms, predominantly non-homologous end joining (NHEJ), act as critical steps to endogenous gene editing or correction. However, there is very limited control over these mechanisms, which generate an abundance of random insertions and deletions (indels). Frameshift mutations associated with these indels of unspecified size and sequence might result in loss of function phenotypes of agronomic importance. Gain of function mutations, on the other hand, generally requires precise editing of the target locus with the aid of a homologous template and involves the cellular homology directed repair (HDR) mechanism. We will present an HDR mediated genome editing approach in sugarcane and will discuss the impact of alternative guide RNAâ€™s and delivered amount of components.

Regulation of Stomatal Movement Using Synthetic Biology Approach

- Conference: International Conference on Plant Synthetic Biology and Bioengineering
- Year: 2016
- Proceeding: International Conference on Plant Synthetic Biology and Bioengineering
- Group: Poster Submissions
- Session:

  Poster Session

- Time:

  Friday, December 16, 2016 - 5:20pm-7:20pm

Authors:
Liu, D., Oak Ridge National Laboratory
Hu, R., Oak Ridge National Laboratory
Borland, A., Newcastle University
Yang, X., Oak Ridge National Laboratory
Crassulacean acid metabolism (CAM) enhances plant water-use efficiency and drought tolerance by reducing transpirational water loss through stomatal closure during the day when the temperature is higher and stomatal opening during the night when the temperature is lower. Engineering CAM into C_3 plants has the potential to sustain production of food, feed, fiber, and biofuel in hotter and drier climates. A key task of CAM-into-C_3 engineering is to inverse the stomatal behavior of C_3 plants, in which stomata are open during the day and closed at night. Stomatal movement depends on the transport of inorganic ions and organic metabolites across guard cell membranes. The transporters are regulated by various signaling pathways such as abscisic acid (ABA), blue light, CO_2 and Ca^{2+} signaling pathways. Based on the newly-sequenced genome of *Kalanchoe laxiflora* (an obligate CAM species with a genome size of ~250Mb), we identified some candidate genes involved in regulation of stomatal movement in CAM plants. We designed and assembled the several genetic circuits of â€˜CAM-like stomatal

movementâ€™ using synthetic biology approaches, and transferred them into three target C_3 species (i.e., *Arabidopsis*, tobacco, poplar) via *Agrobacterium*-mediated transformation. The effects of these genetic circuits on stomatal movement in the transgenic plants will provide useful information for optimizing the strategy of CAM-into-C_3 engineering. Also, success of this research could facilitate future efforts to engineer other multi-gene traits through synthetic biology approaches.

Stable Expression of the Fungal Effector BEC1019 in Transgenic Barley to Evaluate Host-Pathogen Interactions

- Conference: International Conference on Plant Synthetic Biology and Bioengineering
- Year: 2016
- Proceeding: International Conference on Plant Synthetic Biology and Bioengineering
- Group: Poster Submissions
- Session:

  Poster Session

- Time:

  Friday, December 16, 2016 - 5:20pm-7:20pm

Authors:
Oz, T., University of Florida - IFAS
Xu, W.
Wise, R.
Altpeter, F., University of Florida - IFAS

Powdery mildews, caused by obligate fungal pathogens, are a major threat to cereal grain production worldwide. Effector proteins secreted by these pathogens suppress or induce host processes to promote nutrient acquisition and colonization. As such, effectors serve as optimal probes to decode host signaling pathways. We are focusing on the interaction of barley with the powdery mildew fungus, *Blumeria graminis* f. sp. *hordei* (*Bgh*), to address fundamental questions in host resistance and susceptibility. Among ca. 540 predicted secreted effectors, *Blumeria* effector candidate (*BEC*) *1019* is a single-copy gene encoding a putative, secreted metalloprotease. BEC1019 is evolutionarily conserved among at least 96 other diverse fungi, including obligate plant pathogens, necrotrophic animal pathogens, and free-living non-pathogens. It was recently reported that BEC1019 suppresses host cell death and enhances *Bgh* virulence in barley. We postulate that BEC1019 represents an ancient fungal protein family, members of which have evolved to function as effectors in plant and animal hosts. Here we report overexpression of *BEC1019* in barley cv. Golden Promise via *Agrobacterium*-mediated genetic transformation. Transgenic plants were regenerated following selection with hygromycin. Expression levels of *BEC1019* gene variants with or without a signal peptide (SP) and an affinity purification tag (HPB tag), and production of homozygous transgenic seed progenies are reported. Transgenic plants are currently analyzed with pull-down assay to determine cellular localization and confirm target interactions.

Supported by the National Science Foundation - Plant Genome grant 13-39348.

SynCellBio: An Inquiry-Based Research Course Exploring Synthetic Biology

- Conference: International Conference on Plant Synthetic Biology and Bioengineering
- Year: 2016
- Proceeding: International Conference on Plant Synthetic Biology and Bioengineering
- Group: Poster Submissions
- Session:

  Poster Session

- Time:

  Friday, December 16, 2016 - 5:20pm-7:20pm

Authors:
Moss, B., Whitman College
Synthetic biology is ideally poised for integration within an undergraduate liberal arts curriculum. Topics within synthetic biology, such as genetic engineering and the generation of new lifeforms, rely upon cross-disciplinary approaches from cell biology, engineering, molecular genetics, and biochemistry. Study of these topics readily lead into broader discussions of scientific research regulation, ethics, food security, and politics. Additionally, many of the experimental approaches used in synthetic biology are readily learned and executed by undergraduates, enabling implementation of a hands-on laboratory course component. Thus, I have developed an inquiry-based synthetic biology course (SynCellBio) that utilizes a fused lecture-discussion-laboratory format that meets twice weekly. The course is a cell/molecular biology elective enrolling 14-18 students per semester. The learning goals for SynCellBio are (1) to survey the field of synthetic biology, (2) to engage students in an authentic synthetic biology research experience, and (3) to develop proficiency and breadth communicating scientific research. Students survey seminal findings and current advances in synthetic biology through weekly,

student-led journal club presentations and instructor-directed group activities. In the research component of the course, students set out to fine-tune a molecular component within a synthetic yeast cell signaling system that is based on the plant hormone auxin. Students devise point mutations that they predict will modulate the degradation rate of auxin repressor parts within the synthetic yeast system. Each student writes a short NSF GRFP-style research proposal based on their hypothesis, participates in several rounds of peer review, and serves as a reviewer in a mock grant panel. Students then work in small teams to build their novel auxin repressor components, express them in the synthetic yeast system, and assess their function by fluorescence microscopy and flow cytometry. Final results are presented at a poster session. Subsequent semesters of the course build off student work from previous semesters, with the ultimate goal of publication and dissemination of the parts to the wider synthetic biology community.

Synthetic Biology of N2-Fixing Cyanobacteria for Photosynthetic Production of Perfumed Linalool from Air and Water

- Conference: International Conference on Plant Synthetic Biology and Bioengineering
- Year: 2016
- Proceeding: International Conference on Plant Synthetic Biology and Bioengineering
- Group: Poster Submissions
- Session:

  Poster Session

- Time:

  Friday, December 16, 2016 - 5:20pm-7:20pm

Authors:
Zhou, R., South Dakota State University, Brookings
Gu, L.
Halfmann, C., South Dakota State University
VanDenTop, T., South Dakota State University
Braselton, N., South Dakota State University
Gibbons, W.
Linalool ($C_{10}H_{18}O$) is a naturally-occurring terpene alcohol, emitted as a volatile from many flowers and over 200 aromatic plants. In plants, linalool is produced from the universal isoprenoid intermediate geranyl diphosphate (GPP) through the MEP pathway. Linalool has many commercial applications including perfumed hygiene products, flavor and fragrances, pharmaceuticals, and drop-in biofuels. Commercial production of linalool from plant extracts is limited by availability of raw feedstocks and high extraction costs.

Engineering N_2-fixing cyanobacteria to photosynthetically produce and secrete linalool may have potential to simultaneously solve these two problems.

Unlike plants, N_2-fixing cyanobacteria are capable of performing both solar-powered N_2-fixation and CO_2-fixation. Their simple input requirements for rapid growth and ease of genetic manipulation, together with the highest photosynthetic efficiency and suitability for industrialized production, N_2-fixing cyanobacteria are attractive organisms for engineering production of linalool. Cyanobacteria, like plants, have native metabolic pathways (Calvin cycle and MEP pathway) to photosynthetically convert CO_2 and water into a variety of reduced carbon compounds, including GPP, the precursor for linalool. However, cyanobacteria lack the linalool synthase that plants use to convert GPP into linalool. In this research, we have created three generations of linalool-producing cyanobacteria. First, a linalool synthase gene from *Norway Spruce* was fused to a synthetic, dual cyanobacterial *P_{nir}â€"P_{psbA1}* promoter and subcloned into a shuttle vector for transformation of an N_2-fixing cyanobacterium *Anabaena sp.* PCC7120. **The first generation** of transgenic *Anabaena* (with a single plant linalool synthase gene) was confirmed to continuously synthesize and secrete linalool using only air (CO_2 and N_2 gas), mineralized water and light energy. To increase linalool production, a synthetic operon coding for three key enzymes (DXS-IDI-GPPS) from the MEP pathway were over-expressed in *Anabaena*. **This second generation** strain produced 2.8 fold more linalool than the first generation strain. Under 50 $\mu E \cdot m^{-2} \cdot s^{-1}$ illumination, the maximum linalool productivity was 35µg/L/day/OD_{720} when nitrate was used as the sole nitrogen source, and 36.1µg/L/day/OD_{720} when atmospheric N_2 gas was the sole nitrogen source. To further increase linalool production, we created a **third generation** of the linalool-producing *Anabaena* by introducing a synthetic CO_2-fixing photorespiratory bypass that is based on the 3-hydroxypropionate bi-cycle (3-HP-bicycle). This synthetic photorespiratory bypass route containing six enzymes is expected to avoid the loss of fixed carbon during photorespiration, and also functions as an additional O_2-tolerant, CO_2-fixing pathway, alongside the Calvin cycle. The third generation strain showed a maximum productivity of 133.7µg/L/day/OD_{720} when the culture was supplemented with 1.0% CO_2,producing about10.7-fold more linalool than the first generation strain.

In summary, we used synthetic biology to have developed N_2-fixing cyanobacteria that can directly convert solar energy, CO_2 and water into a fragrant linalool. The linalool produced by the engineered cyanobacterium was secreted into the media and volatilized into the flask headspace, allowing for easy separation of linalool from the culture biomass. Thus, the engineered N_2-fixing cyanobacterium can serve as a living cellular factory to continuously produce and emit a wide range of commodity chemicals/fuels using only atmospheric gases (CO_2 & N_2), mineralized H_2O, and sunlight.

Talen Mediated Multi-Allelic Mutagenesis of COMT Improves Saccharification Efficiency from Lignocellulosic Biomass of Field Grown Sugarcane

ICPSBB
International Conference on Plant Synthetic Biology and Bioengineering

- Conference: International Conference on Plant Synthetic Biology and Bioengineering
- Year: 2016
- Proceeding: International Conference on Plant Synthetic Biology and Bioengineering
- Group: Poster Submissions
- Session:

  Poster Session

- Time:

  Friday, December 16, 2016 - 5:20pm-7:20pm

Authors:
Kannan, B., University of Florida - IFAS
Jung, J. H., University of Florida - IFAS
Altpeter, F., University of Florida - IFAS
Sugarcane is a prime feedstock for commercial production of bioethanol. Utilizing both sucrose and cell wall bound sugars for fermentation will enhance the biofuel yield per unit area. We recently demonstrated that RNAi mediated downregulation of lignin biosynthetic gene Caffeic acid O-methyltransferase (COMT) is a successful strategy to improve bioethanol production from lignocellulosic sugarcane biomass. Targeted mutagenesis with genome editing tools has been successful in a number of crops for creation of loss of function phenotypes but is very challenging in sugarcane due to its highly polyploid genome. In this study, a conserved region of COMT was targeted with a

single TALEN pair for multi-allelic mutagenesis to modify lignin biosynthesis in sugarcane. Targeted mutations in regenerated sugarcane lines were identified by capillary electrophoresis of the COMT amplicon. Mutations were confirmed by high-throughput sequencing of the COMT amplicon which revealed the presence of insertions and deletions at the target region. Long COMT amplicons of vegetative progenies, which combined excellent agronomic performance with altered cell wall composition and drastically improved saccharification efficiency were analyzed with Sanger sequencing. Sequencing revealed the extent of co-editing of the large number of COMT copies/alleles and confirms the success of this strategy in a highly polyploid crop.

Towards Design of Endophytic Bacteria That Can Deliver Fixed Nitrogen to Cereal Crops

- Conference: International Conference on Plant Synthetic Biology and Bioengineering
- Year: 2016
- Proceeding: International Conference on Plant Synthetic Biology and Bioengineering
- Group: Poster Submissions
- Session:

  Poster Session

- Time:

  Friday, December 16, 2016 - 5:20pm-7:20pm

Authors:
Ryu, M. H., MIT
Song, M., MIT
Garcia, K., University of Wisconsin-Madison
Rajasekar, S., University of Wisconsin-Madison
Ané, J. M., University of Wisconsin-Madison
Most cereal crops require a significant amount of nitrogen fertilizer to reach their yield potential. Acquiring nitrogen fertilizers is difficult for poor farmers in developing countries. In contrast, the excessive use of nitrogen fertilizers in developed countries leads to significant environmental and sustainability problems. Unfortunately, unlike legumes, most cereals cannot obtain a significant amount of nitrogen through biological nitrogen fixation. Low oxygen levels (micro-aerobic conditions) and carbon availability are two prerequisites for efficient biological nitrogen fixation with most aerobic bacteria. We hypothesized that these two features may be more available deep inside of root tissues than on their surface. Bacterial endophytes can colonize plant tissues internally without eliciting significant host defense responses. These endophytic bacteria enter root

tissues through cracks caused by lateral root formation or wounds. Thus, we are aimed at designing endophytic bacteria residing in cereal roots such that they are capable of fixing nitrogen efficiently and providing nitrogen to the plant. Preliminary evidence shows the activity of nitrogenase after the colonization of maize with some engineered endophytic bacteria. The control of nitrogen fixation is therefore ready to be placed under synthetic control, for example, of plant-derived signals in chassis organisms.